Introduction

Mathematical Thinking at Grade 4

Grade 4

Also appropriate for Grade 5

Cornelia C. Tierney

Developed at TERC, Cambridge, Massachusetts

Dale Seymour Publications®
Menlo Park, California

The *Investigations* curriculum was developed at TERC (formerly Technical Education Research Centers) in collaboration with Kent State University and the State University of New York at Buffalo. The work was supported in part by National Science Foundation Grant No. ESI-9050210. TERC is a nonprofit company working to improve mathematics and science education. TERC is located at 2067 Massachusetts Avenue, Cambridge, MA 02140.

This project was supported, in part, by the
National Science Foundation
Opinions expressed are those of the authors and not necessarily those of the Foundation

Managing Editor: Catherine Anderson
Series Editor: Beverly Cory
Revision Team: Laura Marshall Alavosus, Ellen Harding, Patty Green Holubar, Suzanne Knott, John Lanyi, Beverly Hersh Lozoff
ESL Consultant: Nancy Sokol Green
Production/Manufacturing Director: Janet Yearian
Production/Manufacturing Manager: Karen Edmonds
Production/Manufacturing Coordinator: Joe Conte
Design Manager: Jeff Kelly
Design: Don Taka
Illustrations: Susan Jaekel, Carl Yoshihara
Cover: Bay Graphics
Composition: Archetype Book Composition

This book is published by Dale Seymour Publications®, an imprint of Addison Wesley Longman, Inc.

Dale Seymour Publications
2725 Sand Hill Road
Menlo Park, CA 94025
Customer Service: 800-872-1100

Order number DS43890
ISBN 1-57232-743-X
1 2 3 4 5 6 7 8 9 10-ML-01 00 99 98 97

Printed on Recycled Paper

TERC

INVESTIGATIONS IN NUMBER, DATA, AND SPACE®

Principal Investigator **Susan Jo Russell**

Co-Principal Investigator **Cornelia C. Tierney**

Director of Research and Evaluation **Jan Mokros**

Curriculum Development

Joan Akers
Michael T. Battista
Mary Berle-Carman
Douglas H. Clements
Karen Economopoulos
Ricardo Nemirovsky
Andee Rubin
Susan Jo Russell
Cornelia C. Tierney
Amy Shulman Weinberg

Evaluation and Assessment

Mary Berle-Carman
Abouali Farmanfarmaian
Jan Mokros
Mark Ogonowski
Amy Shulman Weinberg
Tracey Wright
Lisa Yaffee

Teacher Support

Rebecca B. Corwin
Karen Economopoulos
Tracey Wright
Lisa Yaffee

Technology Development

Michael T. Battista
Douglas H. Clements
Julie Sarama Meredith
Andee Rubin

Video Production

David A. Smith

Cooperating Classrooms for This Unit

Ruth Shea
Cambridge Public Schools
Cambridge, MA

Kathleen O'Connell
Arlington Public Schools
Arlington, MA

Nancy Buell
Brookline Public Schools
Brookline, MA

Consultants and Advisors

Elizabeth Badger
Deborah Lowenberg Ball
Marilyn Burns
Ann Grady
Joanne M. Gurry
James J. Kaput
Steven Leinwand
Mary M. Lindquist
David S. Moore
John Olive
Leslie P. Steffe
Peter Sullivan
Grayson Wheatley
Virginia Woolley
Anne Zarinnia

Administration and Production

Amy Catlin
Amy Taber

Graduate Assistants

Kent State University

Joanne Caniglia
Pam DeLong
Carol King

State University of New York at Buffalo

Rosa Gonzalez
Sue McMillen
Julie Sarama Meredith
Sudha Swaminathan

Revisions and Home Materials

Cathy Miles Grant
Marlene Kliman
Margaret McGaffigan
Megan Murray
Kim O'Neil
Andee Rubin
Susan Jo Russell
Lisa Seyferth
Myriam Steinback
Judy Storeygard
Anna Suarez
Cornelia Tierney
Carol Walker
Tracey Wright

CONTENTS

TEACHER NOTES

WHERE TO START

The first-time user of *Mathematical Thinking at Grade 4* should read the following:

When you next teach this same unit, you can begin to read more of the background. Each time you present the unit, you will learn more about how your students understand the mathematical ideas.

ABOUT THE *INVESTIGATIONS* CURRICULUM

Investigations in Number, Data, and Space® is a K–5 mathematics curriculum with four major goals:

- to offer students meaningful mathematical problems
- to emphasize depth in mathematical thinking rather than superficial exposure to a series of fragmented topics
- to communicate mathematics content and pedagogy to teachers
- to substantially expand the pool of mathematically literate students

The *Investigations* curriculum embodies a new approach based on years of research about how children learn mathematics. Each grade level consists of a set of separate units, each offering 2–8 weeks of work. These units of study are presented through investigations that involve students in the exploration of major mathematical ideas.

Approaching the mathematics content through investigations helps students develop flexibility and confidence in approaching problems, fluency in using mathematical skills and tools to solve problems, and proficiency in evaluating their solutions. Students also build a repertoire of ways to communicate about their mathematical thinking, while their enjoyment and appreciation of mathematics grows.

The investigations are carefully designed to invite all students into mathematics—girls and boys, members of diverse cultural, ethnic, and language groups, and students with different strengths and interests. Problem contexts often call on students to share experiences from their family, culture, or community. The curriculum eliminates barriers—such as work in isolation from peers, or emphasis on speed and memorization—that exclude some students from participating successfully in mathematics. The following aspects of the curriculum ensure that all students are included in significant mathematics learning:

- Students spend time exploring problems in depth.
- They find more than one solution to many of the problems they work on.
- They invent their own strategies and approaches, rather than relying on memorized procedures.
- They choose from a variety of concrete materials and appropriate technology, including calculators, as a natural part of their everyday mathematical work.
- They express their mathematical thinking through drawing, writing, and talking.
- They work in a variety of groupings—as a whole class, individually, in pairs, and in small groups.
- They move around the classroom as they explore the mathematics in their environment and talk with their peers.

While reading and other language activities are typically given a great deal of time and emphasis in elementary classrooms, mathematics often does not get the time it needs. If students are to experience mathematics in depth, they must have enough time to become engaged in real mathematical problems. We believe that a minimum of five hours of mathematics classroom time a week—about an hour a day—is critical at the elementary level. The plan and pacing of the *Investigations* curriculum is based on that belief.

We explain more about the pedagogy and principles that underlie these investigations in Teacher Notes throughout the units. For correlations of the curriculum to the NCTM Standards and further help in using this research-based program for teaching mathematics, see the following books:

- *Implementing the* Investigations in Number, Data, and Space® *Curriculum*
- *Beyond Arithmetic: Changing Mathematics in the Elementary Classroom* by Jan Mokros, Susan Jo Russell, and Karen Economopoulos

This book is one of the curriculum units for *Investigations in Number, Data, and Space.* In addition to providing part of a complete mathematics curriculum for your students, this unit offers information to support your own professional development. You, the teacher, are the person who will make this curriculum come alive in the classroom; the book for each unit is your main support system.

Although the curriculum does not include student textbooks, reproducible sheets for student work are provided in the unit and are also available as Student Activity Booklets. Students work actively with objects and experiences in their own environment and with a variety of manipulative materials and technology, rather than with a book of instruction and problems. We strongly recommend use of the overhead projector as a way to present problems, to focus group discussion, and to help students share ideas and strategies.

Ultimately, every teacher will use these investigations in ways that make sense for his or her particular style, the particular group of students, and the constraints and supports of a particular school environment. Each unit offers information and guidance for a wide variety of situations, drawn from our collaborations with many teachers and students over many years. Our goal in this book is to help you, a professional educator, implement this curriculum in a way that will give all your students access to mathematical power.

Investigation Format

The opening two pages of each investigation help you get ready for the work that follows.

What Happens This gives a synopsis of each session or block of sessions.

Mathematical Emphasis This lists the most important ideas and processes students will encounter in this investigation.

What to Plan Ahead of Time These lists alert you to materials to gather, sheets to duplicate, transparencies to make, and anything else you need to do before starting.

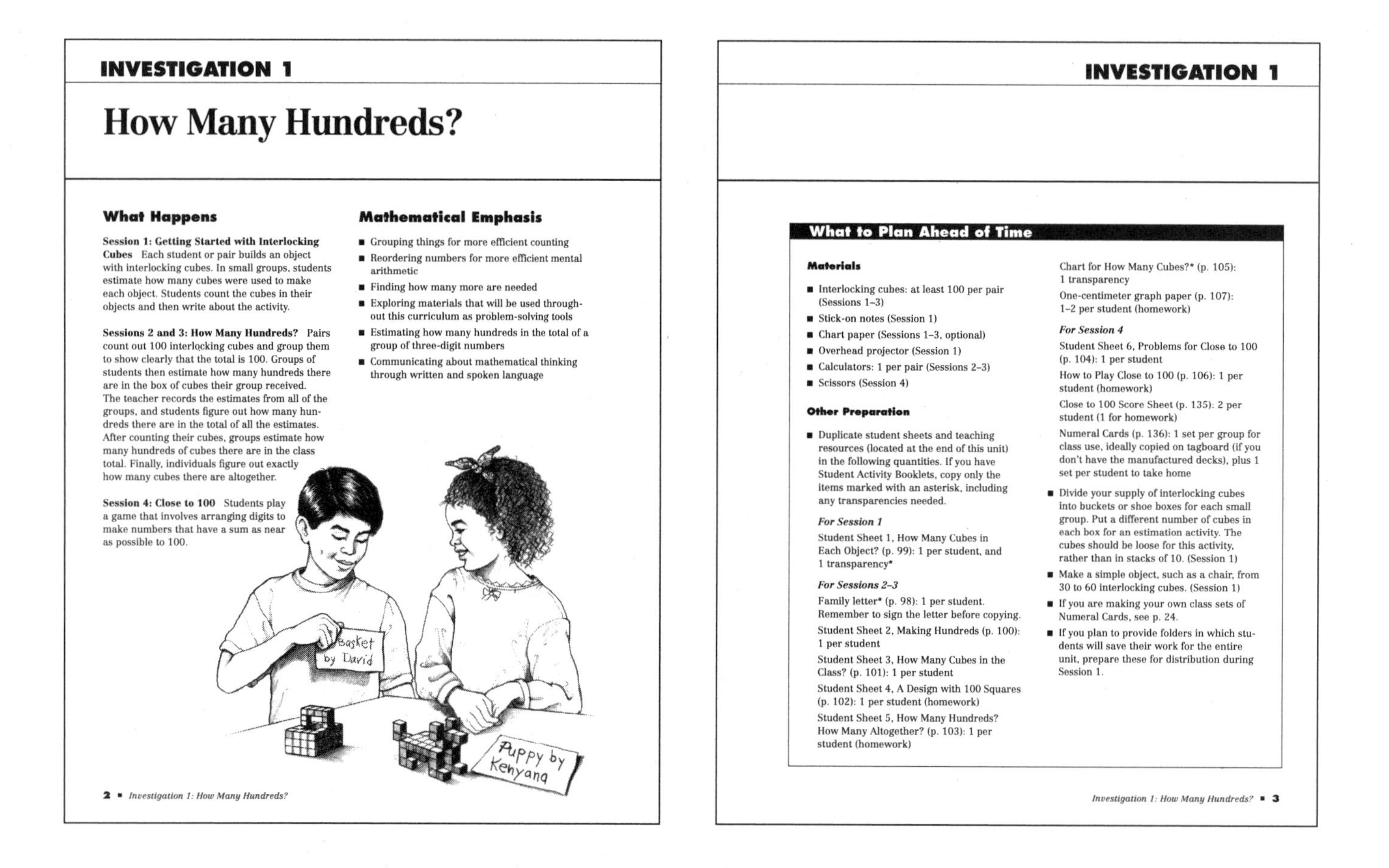

INVESTIGATION 1

How Many Hundreds?

What Happens

Session 1: Getting Started with Interlocking Cubes Each student or pair builds an object with interlocking cubes. In small groups, students estimate how many cubes were used to make each object. Students count the cubes in their objects and then write about the activity.

Sessions 2 and 3: How Many Hundreds? Pairs count out 100 interlocking cubes and group them to show clearly that the total is 100. Groups of students then estimate how many hundreds there are in the box of cubes their group received. The teacher records the estimates from all of the groups, and students figure out how many hundreds there are in the total of all the estimates. After counting their cubes, groups estimate how many hundreds of cubes there are in the class total. Finally, individuals figure out exactly how many cubes there are altogether.

Session 4: Close to 100 Students play a game that involves arranging digits to make numbers that have a sum as near as possible to 100.

Mathematical Emphasis

- Grouping things for more efficient counting
- Reordering numbers for more efficient mental arithmetic
- Finding how many more are needed
- Exploring materials that will be used throughout this curriculum as problem-solving tools
- Estimating how many hundreds in the total of a group of three-digit numbers
- Communicating about mathematical thinking through written and spoken language

2 ▪ Investigation 1: How Many Hundreds?

INVESTIGATION 1

What to Plan Ahead of Time

Materials

- Interlocking cubes: at least 100 per pair (Sessions 1–3)
- Stick-on notes (Session 1)
- Chart paper (Sessions 1–3, optional)
- Overhead projector (Session 1)
- Calculators: 1 per pair (Sessions 2–3)
- Scissors (Session 4)

Other Preparation

- Duplicate student sheets and teaching resources (located at the end of this unit) in the following quantities. If you have Student Activity Booklets, copy only the items marked with an asterisk, including any transparencies needed.

  For Session 1

  Student Sheet 1, How Many Cubes in Each Object? (p. 99): 1 per student, and 1 transparency*

  For Sessions 2–3

  Family letter* (p. 98): 1 per student. Remember to sign the letter before copying.

  Student Sheet 2, Making Hundreds (p. 100): 1 per student

  Student Sheet 3, How Many Cubes in the Class? (p. 101): 1 per student

  Student Sheet 4, A Design with 100 Squares (p. 102): 1 per student (homework)

  Student Sheet 5, How Many Hundreds? How Many Altogether? (p. 103): 1 per student (homework)

  Chart for How Many Cubes?* (p. 105): 1 transparency

  One-centimeter graph paper (p. 107): 1–2 per student (homework)

  For Session 4

  Student Sheet 6, Problems for Close to 100 (p. 104): 1 per student

  How to Play Close to 100 (p. 106): 1 per student (homework)

  Close to 100 Score Sheet (p. 135): 2 per student (1 for homework)

  Numeral Cards (p. 136): 1 set per group for class use, ideally copied on tagboard (if you don't have the manufactured decks), plus 1 set per student to take home

- Divide your supply of interlocking cubes into buckets or shoe boxes for each small group. Put a different number of cubes in each box for an estimation activity. The cubes should be loose for this activity, rather than in stacks of 10. (Session 1)
- Make a simple object, such as a chair, from 30 to 60 interlocking cubes. (Session 1)
- If you are making your own class sets of Numeral Cards, see p. 24.
- If you plan to provide folders in which students will save their work for the entire unit, prepare these for distribution during Session 1.

Investigation 1: How Many Hundreds? ▪ 3

Sessions Within an investigation, the activities are organized by class session, a session being at least a one-hour math class. Sessions are numbered consecutively through an investigation. Often several sessions are grouped together, presenting a block of activities with a single major focus.

When you find a block of sessions presented together—for example, Sessions 1, 2, and 3—read through the entire block first to understand the overall flow and sequence of the activities. Make some preliminary decisions about how you will divide the activities into three sessions for your class, based on what you know about your students. You may need to modify your initial plans as you progress through the activities, and you may want to make notes in the margins of the pages as reminders for the next time you use the unit.

Be sure to read the Session Follow-Up section at the end of the session block to see what homework assignments and extensions are suggested as you make your initial plans.

While you may be used to a curriculum that tells you exactly what each class session should cover, we have found that the teacher is in a better position to make these decisions. Each unit is flexible and may be handled somewhat differently by every teacher. While we provide guidance for how many sessions a particular group of activities is likely to need, we want you to be active in determining an appropriate pace and the best transition points for your class. It is not unusual for a teacher to spend more or less time than is proposed for the activities.

Ten-Minute Math At the beginning of some sessions, you will find Ten-Minute Math activities. These are designed to be used in tandem with the investigations, but not during the math hour. Rather, we hope you will do them whenever you have a spare 10 minutes—maybe before lunch or recess, or at the end of the day.

Ten-Minute Math offers practice in key concepts, but not always those being covered in the unit. For example, in a unit on using data, Ten-Minute Math might revisit geometric activities done earlier in the year. Complete directions for the suggested activities are included at the end of each unit.

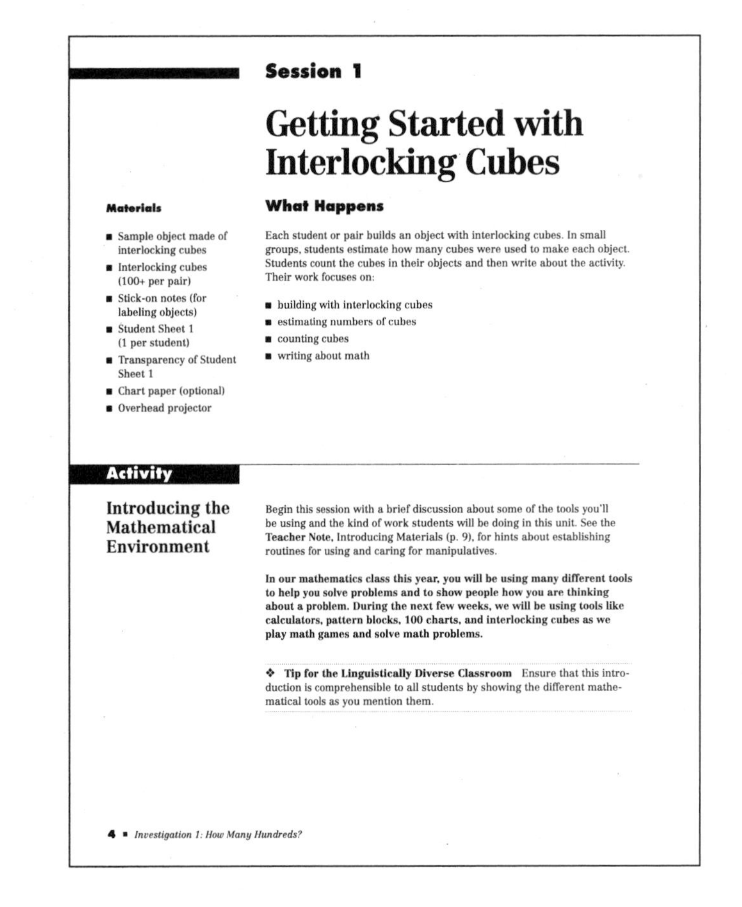
Session 1

Getting Started with Interlocking Cubes

Materials

- Sample object made of interlocking cubes
- Interlocking cubes (100+ per pair)
- Stick-on notes (for labeling objects)
- Student Sheet 1 (1 per student)
- Transparency of Student Sheet 1
- Chart paper (optional)
- Overhead projector

What Happens

Each student or pair builds an object with interlocking cubes. In small groups, students estimate how many cubes were used to make each object. Students count the cubes in their objects and then write about the activity. Their work focuses on:

- building with interlocking cubes
- estimating numbers of cubes
- counting cubes
- writing about math

Activity

Introducing the Mathematical Environment

Begin this session with a brief discussion about some of the tools you'll be using and the kind of work students will be doing in this unit. See the **Teacher Note**, Introducing Materials (p. 9), for hints about establishing routines for using and caring for manipulatives.

In our mathematics class this year, you will be using many different tools to help you solve problems and to show people how you are thinking about a problem. During the next few weeks, we will be using tools like calculators, pattern blocks, 100 charts, and interlocking cubes as we play math games and solve math problems.

❖ **Tip for the Linguistically Diverse Classroom** Ensure that this introduction is comprehensible to all students by showing the different mathematical tools as you mention them.

4 ■ *Investigation 1: How Many Hundreds?*

Activities The activities include pair and small-group work, individual tasks, and whole-class discussions. In any case, students are seated together, talking and sharing ideas during all work times. Students most often work cooperatively, although each student may record work individually.

Choice Time In some units, some sessions are structured with activity choices. In these cases, students may work simultaneously on different activities focused on the same mathematical ideas. Students choose which activities they want to do, and they cycle through them.

You will need to decide how to set up and introduce these activities and how to let students make their choices. Some teachers present them as station activities, in different parts of the room. Some list the choices on the board as reminders or have students keep their own lists.

Extensions Sometimes in Session Follow-Up, you will find suggested extension activities. These are opportunities for some or all students to explore a topic in greater depth or in a different context. They are not designed for "fast" students; mathematics is a multifaceted discipline, and different students will want to go further in different investigations. Look for and encourage the sparks of interest and enthusiasm you see in your students, and use the extensions to help them pursue these interests.

Excursions Some of the *Investigations* units include excursions—blocks of activities that could be omitted without harming the integrity of the unit. This is one way of dealing with the great depth and variety of elementary mathematics—much more than a class has time to explore in any one year. Excursions give you the flexibility to make different choices from year to year, doing the excursion in one unit this time, and next year trying another excursion.

Tips for the Linguistically Diverse Classroom At strategic points in each unit, you will find concrete suggestions for simple modifications of the teaching strategies to encourage the participation of all students. Many of these tips offer alternative ways to elicit critical thinking from students at varying levels of English proficiency, as well as from other students who find it difficult to verbalize their thinking.

The tips are supported by suggestions for specific vocabulary work to help ensure that all students can participate fully in the investigations. The Preview for the Linguistically Diverse Classroom (p. I-22) lists important words that are assumed as part of the working vocabulary of the unit. Second-language learners will need to become familiar with these words in order to understand the problems and activities they will be doing. These terms can be incorporated into students' second-language work before or during the unit. Activities that can be used to present the words are found in the appendix, Vocabulary Support for Second-Language Learners (p. 95). In addition, ideas for making connections to students' language and cultures, included on the Preview page, help the class explore the unit's concepts from a multicultural perspective.

Materials

A complete list of the materials needed for teaching this unit is found on p. I-17. Some of these materials are available in kits for the *Investigations* curriculum. Individual items can also be purchased from school supply dealers.

Classroom Materials In an active mathematics classroom, certain basic materials should be available at all times: interlocking cubes, pencils, unlined paper, graph paper, calculators, things to count with, and measuring tools. Some activities in this curriculum require scissors and glue sticks or tape. Stick-on notes and large paper are also useful materials throughout.

So that students can independently get what they need at any time, they should know where these materials are kept, how they are stored, and how they are to be returned to the storage area. For example, interlocking cubes are best stored in towers of ten; then, whatever the activity, they should be returned to storage in groups of ten at the end of the hour. You'll find that establishing such routines at the beginning of the year is well worth the time and effort.

Technology Calculators are used throughout *Investigations.* Many of the units recommend that you have at least one calculator for each pair. You will find calculator activities, plus Teacher Notes discussing this important mathematical tool, in an early unit at each grade level. It is assumed that calculators will be readily available for student use.

Computer activities at grade 4 use a software program that was developed especially for the *Investigations* curriculum. The program *Geo-Logo*™ is used for activities in the 2-D Geometry unit, *Sunken Ships and Grid Patterns,* where students explore coordinate graphing systems, the use of negative numbers to represent locations in space, and the properties of geometric figures.

How you use the computer activities depends on the number of computers you have available. Suggestions are offered in the geometry units for how to organize different types of computer environments.

Children's Literature Each unit offers a list of suggested children's literature (p. I-17) that can be used to support the mathematical ideas in the unit. Sometimes an activity is based on a specific children's book, with suggestions for substitutions where practical. While such activities can be adapted and taught without the book, the literature offers a rich introduction and should be used whenever possible.

Student Sheets and Teaching Resources Student recording sheets and other teaching tools needed for both class and homework are provided as reproducible blackline masters at the end of each unit. They are also available as Student Activity Booklets. These booklets contain all the sheets each student will need for individual work, freeing you from extensive copying (although you may need or want to copy the occasional teaching resource on transparency film or card stock, or make extra copies of a student sheet).

COIN CARDS (page 1 of 4)

© Dale Seymour Publications® 109 *Investigation 2 • Resource* *Mathematical Thinking at Grade 4*

We think it's important that students find their own ways of organizing and recording their work. They need to learn how to explain their thinking with both drawings and written words, and how to organize their results so someone else can understand them. For this reason, we deliberately do not provide student sheets for every activity. Regardless of the form in which students do their work, we recommend that they keep a mathematics notebook or folder so that their work is always available for reference.

Homework In *Investigations,* homework is an extension of classroom work. Sometimes it offers review and practice of work done in class, sometimes preparation for upcoming activities, and sometimes numerical practice that revisits work in earlier units. Homework plays a role both in supporting students' learning and in helping inform families about the ways in which students in this curriculum work with mathematical ideas.

Depending on your school's homework policies and your own judgment, you may want to assign more homework than is suggested in the units. For this purpose you might use the practice pages, included as blackline masters at the end of this unit, to give students additional work with numbers.

For some homework assignments, you will want to adapt the activity to meet the needs of a variety of students in your class: those with special needs, those ready for more challenge, and second-language learners. You might change the numbers in a problem, make the activity more or less complex, or go through a sample activity with those who need extra help. You can modify any student sheet for either homework or class use. In particular, making numbers in a problem smaller or larger can make the same basic activity appropriate for a wider range of students.

Another issue to consider is how to handle the homework that students bring back to class—how to recognize the work they have done at home without spending too much time on it. Some teachers hold a short group discussion of different approaches to the assignment; others ask students to share and discuss their work with a neighbor, or post the homework around the room and give students time to tour it briefly. If you want to keep track of homework students bring in, be sure it ends up in a designated place.

***Investigations* at Home** It is a good idea to make your policy on homework explicit to both students and their families when you begin teaching with *Investigations*. How frequently will you be assigning homework? When do you expect homework to be completed and brought back to school? What are your goals in assigning homework? How independent should families expect their children to be? What should the parent's or guardian's role be? The more explicit you can be about your expectations, the better the homework experience will be for everyone.

Investigations at Home (a booklet available separately for each unit, to send home with students) gives you a way to communicate with families about the work students are doing in class. This booklet includes a brief description of every session, a list of the mathematics content emphasized in each investigation, and a discussion of each homework assignment to help families more effectively support their children. Whether or not you are using the *Investigations* at Home booklets, we expect you to make your own choices about homework assignments. Feel free to omit any and to add extra ones you think are appropriate.

Family Letter A letter that you can send home to students' families is included with the blackline masters for each unit. Families need to be informed about the mathematics work in your classroom; they should be encouraged to participate in and support their children's work. A reminder to send home the letter for each unit appears in one of the early investigations. These letters are also available separately in Spanish, Vietnamese, Cantonese, Hmong, and Cambodian.

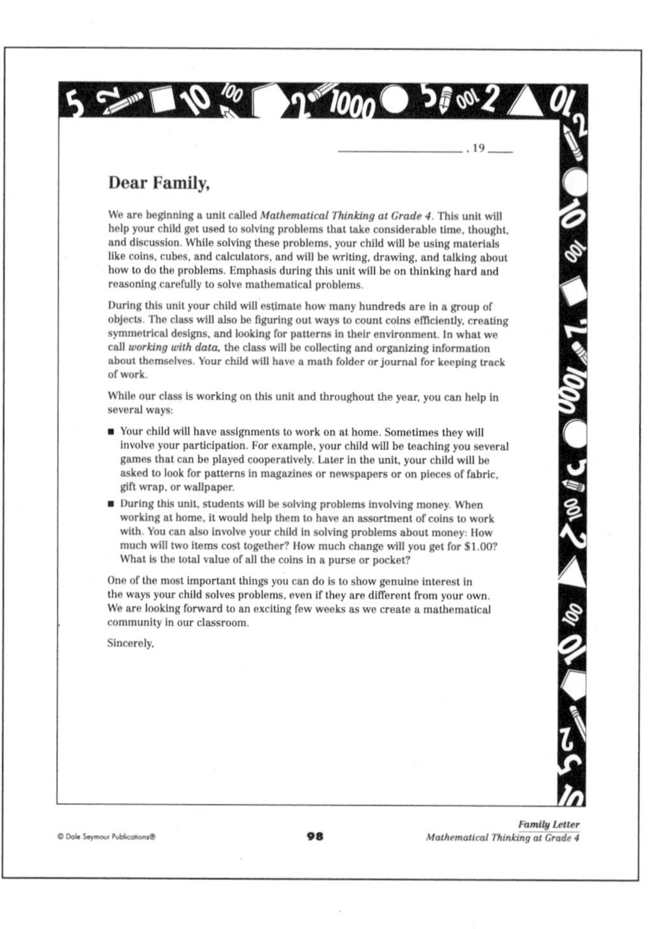

______________, 19____

Dear Family,

We are beginning a unit called *Mathematical Thinking at Grade 4*. This unit will help your child get used to solving problems that take considerable time, thought, and discussion. While solving these problems, your child will be using materials like coins, cubes, and calculators, and will be writing, drawing, and talking about how to do the problems. Emphasis during this unit will be on thinking hard and reasoning carefully to solve mathematical problems.

During this unit your child will estimate how many hundreds are in a group of objects. The class will also be figuring out ways to count coins efficiently, creating symmetrical designs, and looking for patterns in their environment. In what we call *working with data*, the class will be collecting and organizing information about themselves. Your child will have a math folder or journal for keeping track of work.

While our class is working on this unit and throughout the year, you can help in several ways:

- Your child will have assignments to work on at home. Sometimes they will involve your participation. For example, your child will be teaching you several games that can be played cooperatively. Later in the unit, your child will be asked to look for patterns in magazines or newspapers or on pieces of fabric, gift wrap, or wallpaper.
- During this unit, students will be solving problems involving money. When working at home, it would help them to have an assortment of coins to work with. You can also involve your child in solving problems about money: How much will two items cost together? How much change will you get for $1.00? What is the total value of all the coins in a purse or pocket?

One of the most important things you can do is to show genuine interest in the ways your child solves problems, even if they are different from your own. We are looking forward to an exciting few weeks as we create a mathematical community in our classroom.

Sincerely,

Help for You, the Teacher

Because we believe strongly that a new curriculum must help teachers think in new ways about mathematics and about their students' mathematical thinking processes, we have included a great deal of material to help you learn more about both.

About the Mathematics in This Unit This introductory section (p. I-18) summarizes the critical information about the mathematics you will be teaching. It describes the unit's central mathematical ideas and how students will encounter them through the unit's activities.

Teacher Notes These reference notes provide practical information about the mathematics you are teaching and about our experience with how students learn. Many of the notes were written in response to actual questions from teachers, or to discuss important things we saw happening in the field-test classrooms. Some teachers like to read them all before starting the unit, then review them as they come up in particular investigations.

Dialogue Boxes Sample dialogues demonstrate how students typically express their mathematical ideas, what issues and confusions arise in their thinking, and how some teachers have guided class discussions.

These dialogues are based on the extensive classroom testing of this curriculum; many are word-for-word transcriptions of recorded class discussions. They are not always easy reading; sometimes it may take some effort to unravel what the students are trying to say. But this is the value of these dialogues; they offer good clues to how your students may develop and express their approaches and strategies, helping you prepare for your own class discussions.

Where to Start You may not have time to read everything the first time you use this unit. As a first-time user, you will likely focus on understanding the activities and working them out with your students. Read completely through each investigation before starting to present it. Also read those sections listed in the Contents under the heading Where to Start (p. vi).

Pattern and Symmetry

Teacher Note

A pattern creates a regularity that can be copied. When we speak of patterns in fabric, we may mean a design made from colors or textures. Plaids have a color pattern; corduroys have a texture pattern. A pattern in a mathematical sense has a relationship among its component parts—sometimes in number, sometimes in relationship among data, sometimes in visual shape.

Number A pattern exists in each of these series of number relationships, making it possible to predict the next steps:

10 × 23 = 230	5 + 8 = 13
100 × 23 = 2300	5 + 18 = 23
1,000 × 23 = 23,000	5 + 28 = 33
10,000 × 23 = ?	5 + 38 = ?
? = ?	? = ?

1 3 6 10 ?

Data A fourth grade class that collected bedtime and rising time data in their school concluded that younger children sleep for more hours than older children. Although this pattern had some remarkable exceptions—such as one first grader who reported to sleep only five and a half hours a night—the information allowed them to make generalizations about the number of hours of sleep required by students at different ages, and to make predictions about the sleep patterns of children younger and older than those in their survey.

Shape The designs that follow have patterns that allow you to know what piece is missing. Patterns in which one half can be flipped or folded over onto the other half have mirror symmetry. The two halves are reflections of one another. The first three designs have mirror symmetry (if the question mark in each is replaced with the appropriate shape).

Mirror symmetry

Patterns that revolve around a central point have *rotational symmetry* or *circular symmetry*. A pattern with rotational symmetry can be turned a fraction of a circle—one-third, or 120° if it has three segments; one-fourth, or 90° if it has four segments; one-sixth, or 60° if it has six segments; one-twelfth, or 30° if it has twelve segments—and appear as though it hasn't moved at all. The two patterns below have rotational symmetry.

Only rotational symmetry

Mirror and rotational symmetry

A circle has infinite rotational symmetry; any turn will leave it looking the same. Some patterns with rotational symmetry—such as a circle, a regular hexagon like the yellow pattern block, any even-sided regular polygon, or the six-pointed star above—also have mirror symmetry. Others, such as the pinwheel design above and the trapezoid below, do not.

Mirror and rotational symmetry

Only mirror symmetry

Session 1: Patterns with Mirror Symmetry ▪ **71**

DIALOGUE BOX

Discussing Symmetry

As the students in one class brought in from home their observations of things that have symmetry, the teacher took some time to discuss what they were finding.

Tell us some of the symmetrical things on your list that you saw but couldn't bring to class.

Kenyana: The smoke detector. It is round and you can cut it a few different ways and get it to be a mirror reflection.

B.J.: The curtains.

So there is one curtain on each side of the window, and they match each other?

B.J.: Yes.

Vanessa: A chair, but it matters how you split it. It is symmetrical only if you split the back and seat down the middle one way.

Marci: My cat.

Nick: No living thing is. I thought my dog was, but he's not because he has more spots on one side.

Marci: My cat is exactly the same on both sides.

Elena: The clock is symmetrical.

Is the clock symmetrical?

Dave: At 6 o'clock it is.

Marci: What about the second hand?

Vanessa: I don't think it is because of the numbers. If you took the numbers off, it might be symmetrical.

Marci: You'd have to take the hands off too. They are different lengths, and the second hand really gets in the way.

Does everybody agree that if we took the hands and numbers off, the clock would be symmetrical?

Nick: I have to think about it. It's hard for me to imagine it that way.

While you think about that, I have another question for you. If there is such a thing as a symmetrical cat, how many lines of symmetry does it have? How many ways could you cut it in half and have the same image on each side?

Tyrone: Only one way. You'd have to cut it between the ears and the long way down the tail.

That is an example of mirror symmetry. What else have we talked about that has mirror symmetry?

Marci: The chair, and probably the curtains.

Does anybody know what it means to rotate?

Nadim: To turn around?

I agree with that. Are there any items we mentioned that if you turn them a little bit you can get the same view as when you started? That is, are there any items that have rotational symmetry?

Kenyana: I think the smoke detector is like that. You can cut it a lot of different ways.

Marci: Maybe the clock without numbers and hands.

Tyrone: The chair?

If you were looking at a chair from above, how far would you have to rotate it to get the same view?

Tyrone: All the way around?

Nhat: A full circle.

An object has rotational symmetry when you can rotate it part way and it still looks the same. If you have to rotate it a full circle to get the same view, it doesn't have rotational symmetry.

Tyrone: But it might have mirror symmetry, right?

Session 2: Patterns with Rotational Symmetry ▪ **77**

The *Investigations* curriculum incorporates the use of two forms of technology in the classroom: calculators and computers. Calculators are assumed to be standard classroom materials, available for student use in any unit. Computers are explicitly linked to one or more units at each grade level; they are used with the unit on 2-D geometry at each grade, as well as with some of the units on measuring, data, and changes.

Using Calculators

In this curriculum, calculators are considered tools for doing mathematics, similar to pattern blocks or interlocking cubes. Just as with other tools, students must learn both *how* to use calculators correctly and *when* they are appropriate to use. This knowledge is crucial for daily life, as calculators are now a standard way of handling numerical operations, both at work and at home.

Using a calculator correctly is not a simple task; it depends on a good knowledge of the four operations and of the number system, so that students can select suitable calculations and also determine what a reasonable result would be. These skills are the basis of any work with numbers, whether or not a calculator is involved.

Unfortunately, calculators are often seen as tools to check computations with, as if other methods are somehow more fallible. Students need to understand that any computational method can be used to check any other; it's just as easy to make a mistake on the calculator as it is to make a mistake on paper or with mental arithmetic. Throughout this curriculum, we encourage students to solve computation problems in more than one way in order to double-check their accuracy. We present mental arithmetic, paper-and-pencil computation, and calculators as three possible approaches.

In this curriculum we also recognize that, despite their importance, calculators are not always appropriate in mathematics instruction. Like any tools, calculators are useful for some tasks, but not for others. You will need to make decisions about when to allow students access to calculators and when to ask that they solve problems without them, so that they can concentrate on other tools and skills. At times when calculators are or are not appropriate for a particular activity, we make specific recommendations. Help your students develop their own sense of which problems they can tackle with their own reasoning and which ones might be better solved with a combination of their own reasoning and the calculator.

Managing calculators in your classroom so that they are a tool, and not a distraction, requires some planning. When calculators are first introduced, students often want to use them for everything, even problems that can be solved quite simply by other methods. However, once the novelty wears off, students are just as interested in developing their own strategies, especially when these strategies are emphasized and valued in the classroom. Over time, students will come to recognize the ease and value of solving problems mentally, with paper and pencil, or with manipulatives, while also understanding the power of the calculator to facilitate work with larger numbers.

Experience shows that if calculators are available only occasionally, students become excited and distracted when they are permitted to use them. They focus on the tool rather than on the mathematics. In order to learn when calculators are appropriate and when they are not, students must have easy access to them and use them routinely in their work.

If you have a calculator for each student, and if you think your students can accept the responsibility, you might allow them to keep their calculators with the rest of their individual materials, at least for the first few weeks of school. Alternatively, you might store them in boxes on a shelf, number each calculator, and assign a corresponding number to each student. This system can give students a sense of ownership while also helping you keep track of the calculators.

Using Computers

Students can use computers to approach and visualize mathematical situations in new ways. The computer allows students to construct and manipulate geometric shapes, see objects move according to rules they specify, and turn, flip, and repeat a pattern.

This curriculum calls for computers in units where they are a particularly effective tool for learning mathematics content. One unit on 2-D geometry at each of the grades 3–5 includes a core of activities that rely on access to computers, either in the classroom or in a lab. Other units on geometry, measurement, data, and changes include computer activities, but can be taught without them. In these units, however, students' experience is greatly enhanced by computer use.

The following list outlines the recommended use of computers in this curriculum:

Grade 1
Unit: *Survey Questions and Secret Rules* (Collecting and Sorting Data)
Software: Tabletop, Jr.
Source: Broderbund

Unit: *Quilt Squares and Block Towns* (2-D and 3-D Geometry)
Software: *Shapes*
Source: provided with the unit

Grade 2
Unit: *Mathematical Thinking at Grade 2* (Introduction)
Software: *Shapes*
Source: provided with the unit

Unit: *Shapes, Halves, and Symmetry* (Geometry and Fractions)
Software: *Shapes*
Source: provided with the unit

Unit: *How Long? How Far?* (Measuring)
Software: *Geo-Logo*
Source: provided with the unit

Grade 3
Unit: *Flips, Turns, and Area* (2-D Geometry)
Software: *Tumbling Tetrominoes*
Source: provided with the unit

Unit: *Turtle Paths* (2-D Geometry)
Software: *Geo-Logo*
Source: provided with the unit

Grade 4
Unit: *Sunken Ships and Grid Patterns* (2-D Geometry)
Software: *Geo-Logo*
Source: provided with the unit

Grade 5
Unit: *Picturing Polygons* (2-D Geometry)
Software: *Geo-Logo*
Source: provided with the unit

Unit: *Patterns of Change* (Tables and Graphs)
Software: *Trips*
Source: provided with the unit

Unit: *Data: Kids, Cats, and Ads* (Statistics)
Software: Tabletop, Sr.
Source: Broderbund

The software provided with the *Investigations* units uses the power of the computer to help students explore mathematical ideas and relationships that cannot be explored in the same way with physical materials. With the *Shapes* (grades 1–2) and *Tumbling Tetrominoes* (grade 3) software, students explore symmetry, pattern, rotation and reflection, area, and characteristics of 2-D shapes. With the *Geo-Logo* software (grades 3–5), students investigate rotations and reflections, coordinate geometry, the properties of 2-D shapes, and angles. The *Trips* software (grade 5) is a mathematical exploration of motion in which students run experiments and interpret data presented in graphs and tables.

We suggest that students work in pairs on the computer; this not only maximizes computer resources but also encourages students to consult, monitor, and teach one another. Generally, more than two students at one computer find it difficult to share. Managing access to computers is an issue for every classroom. The curriculum gives you explicit support for setting up a system. The units are structured on the assumption that you have enough computers for half your students to work on the machines in pairs at one time. If you do not have access to that many computers, suggestions are made for structuring class time to use the unit with five to eight computers, or even with fewer than five.

Assessment plays a critical role in teaching and learning, and it is an integral part of the *Investigations* curriculum. For a teacher using these units, assessment is an ongoing process. You observe students' discussions and explanations of their strategies on a daily basis and examine their work as it evolves. While students are busy recording and representing their work, working on projects, sharing with partners, and playing mathematical games, you have many opportunities to observe their mathematical thinking. What you learn through observation guides your decisions about how to proceed. In any of the units, you will repeatedly consider questions like these:

- Do students come up with their own strategies for solving problems, or do they expect others to tell them what to do? What do their strategies reveal about their mathematical understanding?
- Do students understand that there are different strategies for solving problems? Do they articulate their strategies and try to understand other students' strategies?
- How effectively do students use materials as tools to help with their mathematical work?
- Do students have effective ideas for keeping track of and recording their work? Does keeping track of and recording their work seem difficult for them?

You will need to develop a comfortable and efficient system for recording and keeping track of your observations. Some teachers keep a clipboard handy and jot notes on a class list or on adhesive labels that are later transferred to student files. Others keep loose-leaf notebooks with a page for each student and make weekly notes about what they have observed in class.

Assessment Tools in the Unit

With the activities in each unit, you will find questions to guide your thinking while observing the students at work. You will also find two built-in assessment tools: Teacher Checkpoints and embedded Assessment activities.

Teacher Checkpoints The designated Teacher Checkpoints in each unit offer a time to "check in" with individual students, watch them at work, and ask questions that illuminate how they are thinking.

At first it may be hard to know what to look for, hard to know what kinds of questions to ask. Students may be reluctant to talk; they may not be accustomed to having the teacher ask them about their work, or they may not know how to explain their thinking. Two important ingredients of this process are asking students open-ended questions about their work and showing genuine interest in how they are approaching the task. When students see that you are interested in their thinking and are counting on them to come up with their own ways of solving problems, they may surprise you with the depth of their understanding.

Teacher Checkpoints also give you the chance to pause in the teaching sequence and reflect on how your class is doing overall. Think about whether you need to adjust your pacing: Are most students fluent with strategies for solving a particular kind of problem? Are they just starting to formulate good strategies? Or are they still struggling with how to start? Depending on what you see as the students work, you may want to spend more time on similar problems, change some of the problems to use smaller numbers, move quickly to more challenging material, modify subsequent activities for some students, work on particular ideas with a small group, or pair students who have good strategies with those who are having more difficulty.

Embedded Assessment Activities Assessment activities embedded in each unit will help you examine specific pieces of student work, figure out what it means, and provide feedback. From the students' point of view, these assessment activities are no different from any others. Each is a learning experience in and of itself, as well as an opportunity for you to gather evidence about students' mathematical understanding.

The embedded assessment activities sometimes involve writing and reflecting; at other times, a discussion or brief interaction between student and teacher; and in still other instances, the creation and explanation of a product. In most cases, the assessments require that students *show* what they did, *write* or *talk* about it, or do both. Having to explain how they worked through a problem helps students be more focused and clear in their mathematical thinking. It also helps them realize that doing mathematics is a process that may involve tentative starts, revising one's approach, taking different paths, and working through ideas.

Teachers often find the hardest part of assessment to be interpreting their students' work. We provide guidelines to help with that interpretation. If you have used a process approach to teaching writing, the assessment in *Investigations* will seem familiar. For many of the assessment activities, a Teacher Note provides examples of student work and a commentary on what it indicates about student thinking.

Documentation of Student Growth

To form an overall picture of mathematical progress, it is important to document each student's work in journals, notebooks, or portfolios. The choice is largely a matter of personal preference; some teachers have students keep a notebook or folder for each unit, while others prefer one mathematics notebook, or a portfolio of selected work for the entire year. The final activity in each *Investigations* unit, called Choosing Student Work to Save, helps you and the students select representative samples for a record of their work.

This kind of regular documentation helps you synthesize information about each student as a mathematical learner. From different pieces of evidence, you can put together the big picture. This synthesis will be invaluable in thinking about where to go next with a particular child, deciding where more work is needed, or explaining to parents (or other teachers) how a child is doing.

If you use portfolios, you need to collect a good balance of work, yet avoid being swamped with an overwhelming amount of paper. Following are some tips for effective portfolios:

- Collect a representative sample of work, including some pieces that students themselves select for inclusion in the portfolio. There should be just a few pieces for each unit, showing different kinds of work—some assignments that involve writing, as well as some that do not.
- If students do not date their work, do so yourself so that you can reconstruct the order in which pieces were done.
- Include your reflections on the work. When you are looking back over the whole year, such comments are reminders of what seemed especially interesting about a particular piece; they can also be helpful to other teachers and to parents. Older students should be encouraged to write their own reflections about their work.

Assessment Overview

There are two places to turn for a preview of the assessment opportunities in each *Investigations* unit. The Assessment Resources column in the unit Overview Chart (pp. I-13–I-16) identifies the Teacher Checkpoints and Assessment activities embedded in each investigation, guidelines for observing the students that appear within classroom activities, and any Teacher Notes and Dialogue Boxes that explain what to look for and what types of student responses you might expect to see in your classroom. Additionally, the section About the Assessment in This Unit (p. I-20) gives you a detailed list of questions for each investigation, keyed to the mathematical emphases, to help you observe student growth.

Depending on your situation, you may want to provide additional assessment opportunities. Most of the investigations lend themselves to more frequent assessment, simply by having students do more writing and recording while they are working.

Mathematical Thinking at Grade 4

Content of This Unit This unit is meant to familiarize you and your students with the mathematics content and approaches of *Investigations* and to help you assess the strengths and needs of your new class of students. It is not designed to provide a final encounter with important mathematical ideas. Students will revisit key ideas in this unit in greater depth as the year goes on. Even if some of your students have only a partial understanding of some of the topics, we recommend that you not spend more time on this unit than is suggested in the Unit Overview.

The emphases in this unit, and throughout the year, are on mathematical thinking and reasoning, using a variety of tools and models to explore mathematics, and being able to communicate about mathematical ideas through drawing, writing, and talking. Students become familiar with using and caring for materials: interlocking cubes, calculators, play money, geoboards, pattern blocks. They have experiences working cooperatively, recording their work in a systematic way, reporting what they have learned, and writing about their thinking. Tasks are provided for assessing students in number work and geometric thinking.

Connections with Other Units If you are doing the full-year *Investigations* curriculum in the suggested sequence for grade 4, this is the first of eleven units. It has connections with every other unit in the fourth grade sequence, both in its content and in its emphasis on ways of thinking and doing mathematics. The number work in this unit is specifically continued and extended in *Landmarks in the Thousands,* and in the Addition and Subtraction unit, *Money, Miles, and Large Numbers.*

This unit can be used successfully at either grade 4 or grade 5 at any time of the year, depending on the previous experience and needs of your students. It offers a way to help students focus on thinking, working, and talking mathematically, and helps you assess student understanding of some key mathematical content.

Investigations Curriculum ■ Suggested Grade 4 Sequence

▶ *Mathematical Thinking at Grade 4* (Introduction)

Arrays and Shares (Multiplication and Division)

Seeing Solids and Silhouettes (3-D Geometry)

Landmarks in the Thousands (The Number System)

Different Shapes, Equal Pieces (Fractions and Area)

The Shape of the Data (Statistics)

Money, Miles, and Large Numbers (Addition and Subtraction)

Changes Over Time (Graphs)

Packages and Groups (Multiplication and Division)

Sunken Ships and Grid Patterns (2-D Geometry)

Three out of Four Like Spaghetti (Data and Fractions)

Investigation 1 ▪ How Many Hundreds?

Class Sessions	Activities	Pacing
Session 1 (p. 4) GETTING STARTED WITH INTERLOCKING CUBES	Introducing the Mathematical Environment Building with Cubes How Many in Each Object? Writing About the First Session	minimum 1 hr
Sessions 2 and 3 (p. 11) HOW MANY HUNDREDS?	Counting Out 100 Cubes Estimating How Many Cubes Altogether Counting How Many Altogether How Many Hundreds? Finding How Many in All Homework: A Design with 100 Squares Homework: How Many?	minimum 2 hr
Session 4 (p. 23) CLOSE TO 100	Introducing Close to 100 Playing Close to 100 Teacher Checkpoint: Thinking About Close to 100 Homework: Playing Close to 100	minimum 1 hr

Ten-Minute Math ▪ Estimation and Number Sense

Mathematical Emphasis

- Grouping things for more efficient counting
- Recording numbers for more efficient mental arithmetic
- Finding how many more are needed
- Estimating how many hundreds in the total of a group of three-digit numbers
- Communicating about mathematical thinking through written and spoken language
- Exploring materials that will be used throughout this curriculum as problem-solving tools

Assessment Resources

Observing the Students (p. 18)

Introducing Calculators (Teacher Note, p. 20)

Teacher Checkpoint: Thinking About Close to 100 (p. 26)

Strategies for Close to 100 (Dialogue Box, p. 31)

Materials

Interlocking cubes
Stick-on notes
Chart paper
Overhead projector
Calculators
Numeral Cards
Scissors
Student Sheets 1–6
Teaching resource sheets
Family letter

Investigation 2 ■ How Many Dollars?

Class Sessions	Activities	Pacing
Sessions 1 and 2 (p. 34) HOW MUCH MONEY?	Counting Money Adding Coin Values Teacher Checkpoint: The Collecting Dollars Game Homework: Playing Collecting Dollars	minimum 2 hr
Sessions 3 and 4 (p. 40) NUMBER SENSE AND COINS	The Hidden Coins Game Choice Time: Playing Math Games Homework: Playing Hidden Coins	minimum 2 hr
Ten-Minute Math ■ Estimation and Number Sense		

Mathematical Emphasis

- Grouping coins for more efficient counting
- Recognizing values of U.S. coins
- Recognizing the decimal point on the calculator

Assessment Resources

Teacher Checkpoint: The Collecting Dollars Game (p. 38)

Choice Time: Observing the Students (p. 43)

Materials

Play money

Resealable plastic bags for play money

Scissors

Opaque paper bags of real coins, holding 2 quarters, 3 dimes, 3 nickels, and 5 pennies to make exactly $1

Numeral Cards

Calculators

Student Sheet 7

Teaching resource sheets

Investigation 3 ■ Using Number Patterns

Class Sessions	Activities	Pacing
Sessions 1 and 2 (p. 48) THE 300 CHART	Filling in the 300 Chart How Many Steps? Counting by 10's Playing 101 to 200 Bingo Homework: Playing 101 to 200 Bingo	minimum 2 hr
Session 3 (p. 54) RELATED PROBLEM SETS	Solving Sets of Related Problems Teacher Checkpoint: Writing About Strategies Comparing Solutions in Groups Homework: Related Problem Sets	minimum 1 hr
Sessions 4 and 5 (p. 58) ADDITION AND SUBTRACTION STRATEGIES	Choice Time: Learning Together Assessment: Numbers and Money Homework: Complete the Booklet Extension: Invent a New Game	minimum 2 hr
Ten-Minute Math ■ Exploring Data		

Mathematical Emphasis

- Using known answers to find others
- Subtracting on a 300 chart and with a calculator
- Adding and subtracting multiples of ten

Assessment Resources

Filling in the 300 Chart (p. 49)

Teacher Checkpoint: Writing About Strategies (p. 55)

Choice Time: Observing the Students (p. 60)

Assessment: Numbers and Money (Teacher Note, p. 62)

Materials

Calculator

Scissors, tape

Numeral Cards from previous investigations

Game markers (such as cubes, square tiles, counting chips)

Colored pencils, crayons, or markers

Buckets or boxes of interlocking cubes

Alexander, Who Used to Be Rich Last Sunday, by Judith Viorst

Overhead projector

Student Sheets 8–10

Teaching resource sheets

Investigation 4 ■ Making Geometric Patterns

Class Sessions	Activities	Pacing
Session 1 (p. 66) PATTERNS WITH MIRROR SYMMETRY	Symmetry with Pattern Blocks Symmetry in the Environment Homework: Finding Examples of Symmetry Homework: Making a New Design Extension: Making Symmetrical Linear and 3-D Designs	minimum 1 hr
Session 2 (p. 72) PATTERNS WITH ROTATIONAL SYMMETRY	Two Symmetries: Which Is Which? Patterns Around Hexagons What Is a Pattern? Homework: Collecting Designs for Display Extension: Pattern Puzzles	minimum 1 hr
Sessions 3 and 4 (p. 78) PATTERNS AND NONPATTERNS	Teacher Checkpoint: A Display of Patterns Writing About Our Designs Draw the Hidden Design Guessing from Descriptions	minimum 2 hr
Sessions 5 and 6 (p. 83) SYMMETRICAL GEOBOARD PATTERNS	Teacher Checkpoint: Symmetry on the Geoboard Counting Lines of Symmetry Assessment: Mirror and Rotational Symmetry Choosing Student Work to Save Homework: Multiple Lines of Symmetry Extension: Displaying More Patterns	minimum 2 hr

Ten-Minute Math ■ Exploring Data

Mathematical Emphasis

- Distinguishing between geometric patterns and random designs
- Distinguishing between mirror symmetry and rotational symmetry
- Writing about designs

Assessment Resources

Discussing Symmetry (Dialogue Box, p. 77)

Teacher Checkpoint: A Display of Patterns (p. 78)

Teacher Checkpoint: Symmetry on the Geoboard (p. 83)

Assessment: Mirror and Rotational Symmetry (p. 87)

Choosing Student Work to Save (p. 88)

Materials

Pattern blocks
Ruler or straightedge
Pencils, crayons, or markers in red, green, yellow, and blue
Scraps of patterned fabric, wrapping paper, or wallpaper (opt.)
Scissors
Stick-on notes
Geoboards with rubber bands in assorted colors
Overhead projector
Overhead pattern blocks (opt.)
Student Sheets 11–12
Teaching resource sheets

MATERIALS LIST

Following are the basic materials needed for the activities in this unit. Many of the items can be purchased from the publisher, either individually or in the Teacher Resource Package and the Student Materials Kit for grade 4. Detailed information is available on the *Investigations* order form. To obtain this form, call toll-free 1-800-872-1100 and ask for a Dale Seymour customer service representative.

Snap™ Cubes (interlocking cubes): 1 container per 4–6 students (60 per student)

Pattern blocks: 1 bucket per 4–6 students

Overhead pattern blocks (optional)

Play money: at least 2 one-dollar bills, 1 five-dollar bill, 2 fifty-cent pieces, 6 quarters, 8 dimes, 8 nickels, and 10 pennies per pair of students

Real coins: several collections of $1 in 2 quarters, 3 dimes, 3 nickels, and 5 pennies, each in a small paper bag

Numeral Cards (manufactured; or use blackline masters to make your own sets)

Geoboards with rubber bands: 1 per pair

Alexander, Who Used to Be Rich Last Sunday, by Judith Viorst (optional)

Scissors: 1 per student

Calculators: 1 per student

Resealable plastic bags or envelopes for storage of card sets and play money: 3 per student

Old magazines, linoleum brochures, scraps of wrapping paper, scraps of wallpaper, and other sources of patterned pictures and designs (optional)

Overhead projector, pens, and blank transparencies

Stick-on notes

Chart paper

Game markers, such as cubes, square tiles, counting chips: 2 per student

Colored pencils, crayons, or markers

Ruler or straightedge: 1 per pair

The following materials are provided at the end of this unit as blackline masters. A Student Activity Booklet containing all student sheets and teacher resources needed for individual work is available.

Family Letter (p. 98)

Student Sheets 1–12 (p. 99)

Teaching Resources:

- Chart for How Many Cubes? (p. 105)
- How to Play Close to 100 (p. 106)
- One-Centimeter Graph Paper (p. 107)
- Coin Cards (p. 109)
- How to Play Collecting Dollars (p. 113)
- How to Play Hidden Coins (p. 114)
- How to Play 101 to 200 Bingo (p. 126)
- 101 to 200 Bingo Board (p. 127)
- Tens Cards (p. 128)
- Shaded Geoboard Design (p. 132)
- Geoboard Dot Paper (p. 133)
- Triangle Paper (p. 134)
- Close to 100 Score Sheet (p. 135)
- Numeral Cards (p. 136)
- Coin Value Strips (p. 139)
- Choice List (p. 140)

Practice Pages (p. 141)

Related Children's Literature

Dahl, Roald. *Matilda.* New York: Puffin Books, 1988.

Kipling, Rudyard. *The Elephant's Child.* New York: Prentice-Hall, 1987.

Tompert, Ann. *Grandfather Tan's Story.* New York: Crown, 1990.

Viorst, Judith. *Alexander, Who Used to Be Rich Last Sunday.* New York: Macmillan, 1989.

Zimelman, Nathan. *How the Second Grade Got $8205.50 to Visit the Statue of Liberty.* Morton Grove, IL: Albert Whitman, 1992.

Mathematical Thinking at Grade 4, as the title indicates, is designed to be an introduction to mathematical thinking—to some of the content, materials, processes, and ways of working that mathematics entails. Through the work in this unit, we provide experiences to engage students in:

- solving mathematical problems in ways that make sense to them
- talking, writing, and drawing about their work
- working with peers
- building models of mathematical situations
- relying on their own thinking and learning from the thinking of others

Students work with problems in the areas of number, data, and space (geometry). In number they explore what happens when ten or multiples of ten are added or subtracted, work on estimating hundreds in number and whole dollars in sums of money, and begin to develop strategies for combining and comparing large quantities. Their work with geometry includes exploring mirror symmetry and rotational symmetry as they build designs with pattern blocks and on geoboards. In Ten-Minute Math activities related to working with data, students collect information about themselves as a group and begin to find ways of organizing, representing, and discussing the data they have collected. Throughout the unit they are introduced to basic mathematical tools and materials such as interlocking cubes, 100 charts, calculators, money, pattern blocks, and geoboards.

Much of the work in this unit involves patterns. Students work with patterns in a variety of contexts: building with pattern blocks and interlocking cubes, using money, and counting on by 10's and 20's. They look at patterns on the 300 chart and discuss what they know about 100. They develop strategies for picking two-digit numbers with a sum of 100. They distinguish between visual displays that have patterns and those that are pictures or abstract designs without any repeating design.

A major focus of these activities is the development and use of good number sense to combine and compare two-digit and three-digit numbers. We expect that fourth graders know the one-digit addition combinations. Just as common sense grows from experience with the world and how it works, number sense grows from experience with how numbers work. Throughout these activities, students are encouraged to solve addition and subtraction problems by thinking about how the numbers are structured and how they are related to other numbers.

For example, in solving an addition combination they don't know, such as $28 + 38$, students are encouraged to use what they do know—perhaps $30 + 40 - 2 - 2$ or $20 + 30$ and $8 + 8$—to reason about the sum. In solving two- and three-digit addition or subtraction problems, we urge students not to apply rote procedures, but to look at the whole problem first and then apply what they know about the numbers to solve the problem.

Investigation 4 focuses on geometry as a central part of mathematics. Students' investigation of symmetry provides a foundation for their exploration of visual pattern in their environment. Students create designs and color them to highlight symmetry, and find ways to complete designs to create a pattern.

Mathematical Thinking at Grade 4 is designed not only to involve students with some central mathematical concepts but also to introduce them to a particular way of approaching mathematics. Throughout the unit students are encouraged to share their strategies, work cooperatively, use materials, and communicate both verbally and in writing about how they are solving problems. These approaches may be quite difficult for some students. Even taking out, using, and putting away materials may be unfamiliar. Certainly writing and drawing pictures to describe mathematical thinking will be quite difficult for some students.

This unit is a time to focus on the development of these processes; to spend time establishing routines and expectations; to communicate to students your own interest in and respect for their mathematical ideas; to assure students that you want to know about their *thinking,* not just their answers; and to insist that students work hard to solve problems in ways that make sense to them. As the unit unfolds, a mathematical community begins to take shape—a community that you and your students are together responsible for creating and maintaining.

Mathematical Emphasis At the beginning of each investigation, the Mathematical Emphasis section tells you what is most important for students to learn during that investigation. Many of these mathematical understandings and processes are difficult and complex. Students gradually learn more and more about each idea over many years of schooling. Individual students will begin and end the unit with different levels of knowledge and skill, but all will gain greater knowledge about recognizing spatial patterns and solving mathematical problems with number and money in ways that make sense to them.

ABOUT THE ASSESSMENT IN THIS UNIT

Throughout the *Investigations* curriculum, there are many opportunities for ongoing daily assessment as you observe, listen to, and interact with students at work. In this unit, you will find five Teacher Checkpoints:

Investigation 1, Session 4:
Thinking About Close to 100 (p. 26)

Investigation 2, Sessions 1 and 2:
The Collecting Dollars Game (p. 38)

Investigation 3, Session 3:
Writing About Strategies (p. 55)

Investigation 4, Sessions 3 and 4:
A Display of Patterns (p. 78)

Investigation 4, Sessions 5 and 6:
Symmetry on the Geoboard (p. 83)

This unit also has two embedded assessment activities:

Investigation 3, Sessions 4 and 5:
Numbers and Money (p. 61)

Investigation 4, Sessions 5 and 6:
Mirror and Rotational Symmetry (p. 87)

In addition, you can use almost any activity in this unit to assess your students' needs and strengths. Listed below are questions to help you focus your observations in each investigation. You may want to keep track of your observations for each student to help you plan your curriculum and monitor students' growth. Suggestions for documenting student growth can be found on page I-11, in the section About Assessment.

Investigation 1: How Many Hundreds?

- How do students count out 100 objects? Can they count by grouping in more than one way? Can students read and write two-digit and three-digit numbers?
- How do students use landmarks to guide addition and subtraction? For example, do they recognize that 27 + 27 is too small to have a sum close to 100, but that 72 + 26 is close to 100?
- What do students know about the structure of the numbers? For example, do they know that the 80's are closer to 100 than the 40's are; that a number in the 70's requires less than 30 to reach 100; that 723 has 7 hundreds in it, whereas 792 is almost 8 hundreds?
- Do students have a good strategy for solving addition problems involving two-digit and three-digit numbers? Can they break apart and reorder numbers to work on a problem more efficiently?
- How do students use materials and mental strategies to find the difference between two numbers? What strategies do they use to find "how many more"? Are they tied to counting by 1's? Are they frequently off by 1 when comparing two quantities?

Investigation 2: How Many Dollars?

- What strategies do students use to combine coins? Can they count easily by 5's, 10's, 25's? When they add coins, do they need to count on by 1's, or do they recognize how to add 5's and 10's to a number?
- Are students familiar with coins and their values?
- Can students add amounts of money on a calculator, using the decimal point appropriately?

Investigation 3: Using Number Patterns

- How do students use known answers to solve related problems? Do they use their knowledge of place value to see relations among problems?
- How do students work with a 300 chart? Can they add and subtract by tens and hundreds? Do they understand how to use a calculator for subtraction? What strategies do they use for checking the reasonableness of an answer on the calculator?
- Do students know what happens when 10 (or a multiple of 10) is added to or subtracted from a two-digit number and use this to compute efficiently?

Investigation 4: Making Geometric Patterns

- Can students distinguish between a pattern and a design without a pattern?
- Can students make a pattern with mirror symmetry? Can they make a pattern with rotational symmetry? Can students recognize patterns with mirror symmetry? with rotational symmetry? Are students clear about the differences between mirror and rotational symmetry?
- What vocabulary do students use when describing designs? Do students use both words and pictures effectively?

Thinking and Working in Mathematics

Mathematical Thinking at Grade 4 provides the chance for you to observe students' work habits and communication skills. Think about these questions to help you decide which of these routines, processes, and materials will require the most ongoing support, guidance, and opportunities for practice.

- Are students comfortable and focused working together in pairs? In small groups?
- Do students expect to devise their own strategies for solving problems, or do they expect you to tell them what to do, or expect to copy from another student? Do they understand that different people may solve problems in different ways?
- Are students familiar with the basic mathematics materials used in this unit? Do they know how to use them? Do they have strategies to use these materials as tools when solving problems?
- Can students work well with materials? Do they take them out and put them away efficiently?
- Can students express their ideas orally? Who participates in discussions? Are they always the same students? Are there students who never participate?
- Do students have ideas about how to record their work, or does writing and drawing about mathematics seem new to them?
- Can students choose an activity from among several that are offered, then move smoothly to a second activity when finished with the first?

In the *Investigations* curriculum, mathematical vocabulary is introduced naturally during the activities. We don't ask students to learn definitions of new terms; rather, they come to understand such words as *factor* or *area* or *symmetry* by hearing them used frequently in discussion as they investigate new concepts. This approach is compatible with current theories of second-language acquisition, which emphasize the use of new vocabulary in meaningful contexts while students are actively involved with objects, pictures, and physical movement.

Listed below are some key words used in this unit that will not be new to most English speakers at this age level, but may be unfamiliar to students with limited English proficiency. You will want to spend additional time working on these words with your students who are learning English. If your students are working with a second-language teacher, you might enlist your colleague's aid in familiarizing students with these words, before and during this unit. In the classroom, look for opportunities for students to hear and use these words. Activities you can use to present the words are given in the appendix, Vocabulary Support for Second-Language Learners (p. 95).

the numbers 1 to 300 Students add and subtract numbers in the hundreds and use the 300 chart. They should be able to write the numerals and identify each by name.

same, different, compare Students compare many things—two numbers, two patterns, two amounts of money—and determine whether they are the *same* or *different*. These terms are also an important part of checking and double-checking answers to problems.

add, plus, subtract, minus, difference Students perform addition and subtraction throughout the unit, in games and activities, with and without the calculator, and use these terms to describe their work.

lowest, close to In a game they play many times, Close to 100, students need to recognize numbers that are *close to* 100, and need to know what it means to try for the *lowest* score.

money: coins, cents, penny, nickel, dime, quarter, dollar Students need to recognize U.S. coins and know the value of each as they practice counting coins mentally to find the total value.

pattern Students look for patterns on the 300 chart; they find number patterns that can help them solve related computation problems; and they also work with patterns of shape and color as they explore symmetrical designs with pattern blocks.

shape, triangle, square, hexagon, trapezoid, diamond Students use pattern block shapes to build and explore symmetrical patterns.

Multicultural Extensions for All Students

Whenever possible, encourage students to share words, objects, customs, or any aspects of daily life from their own cultures and backgrounds that are relevant to the activities in this unit.

For example, during their exploration of symmetry in Investigation 4, you might help students locate pictures of national flags from around the world (found in almanacs or the encyclopedia) and determine which designs have either mirror or rotational symmetry. Students who have actual flags from their countries of origin might bring them in to share for this discussion. Additionally, students might bring in for display and discussion other emblems, fabrics, or objects with repeated patterns or symmetrical designs.

Investigations

How Many Hundreds?

What Happens

Session 1: Getting Started with Interlocking Cubes Each student or pair builds an object with interlocking cubes. In small groups, students estimate how many cubes were used to make each object. Students count the cubes in their objects and then write about the activity.

Sessions 2 and 3: How Many Hundreds? Pairs count out 100 interlocking cubes and group them to show clearly that the total is 100. Groups of students then estimate how many hundreds there are in the box of cubes their group received. The teacher records the estimates from all of the groups, and students figure out how many hundreds there are in the total of all the estimates. After counting their cubes, groups estimate how many hundreds of cubes there are in the class total. Finally, individuals figure out exactly how many cubes there are altogether.

Session 4: Close to 100 Students play a game that involves arranging digits to make numbers that have a sum as near as possible to 100.

Mathematical Emphasis

- Grouping things for more efficient counting
- Reordering numbers for more efficient mental arithmetic
- Finding how many more are needed
- Exploring materials that will be used throughout this curriculum as problem-solving tools
- Estimating how many hundreds in the total of a group of three-digit numbers
- Communicating about mathematical thinking through written and spoken language

What to Plan Ahead of Time

Materials

- Interlocking cubes: at least 100 per pair (Sessions 1–3)
- Stick-on notes (Session 1)
- Chart paper (Sessions 1–3, optional)
- Overhead projector (Session 1)
- Calculators: 1 per pair (Sessions 2–3)
- Scissors (Session 4)

Other Preparation

- Duplicate student sheets and teaching resources (located at the end of this unit) in the following quantities. If you have Student Activity Booklets, copy only the items marked with an asterisk, including any transparencies needed.

  For Session 1

  Student Sheet 1, How Many Cubes in Each Object? (p. 99): 1 per student, and 1 transparency*

  For Sessions 2–3

  Family letter* (p. 98): 1 per student. Remember to sign the letter before copying.

  Student Sheet 2, Making Hundreds (p. 100): 1 per student

  Student Sheet 3, How Many Cubes in the Class? (p. 101): 1 per student

  Student Sheet 4, A Design with 100 Squares (p. 102): 1 per student (homework)

  Student Sheet 5, How Many Hundreds? How Many Altogether? (p. 103): 1 per student (homework)

  Chart for How Many Cubes?* (p. 105): 1 transparency

  One-centimeter graph paper (p. 107): 1–2 per student (homework)

  For Session 4

  Student Sheet 6, Problems for Close to 100 (p. 104): 1 per student

  How to Play Close to 100 (p. 106): 1 per student (homework)

  Close to 100 Score Sheet (p. 135): 2 per student (1 for homework)

  Numeral Cards (p. 136): 1 set per group for class use, ideally copied on tagboard (if you don't have the manufactured decks), plus 1 set per student to take home

- Divide your supply of interlocking cubes into buckets or shoe boxes for each small group. Put a different number of cubes in each box for an estimation activity. The cubes should be loose for this activity, rather than in stacks of 10. (Session 1)
- Make a simple object, such as a chair, from 30 to 60 interlocking cubes. (Session 1)
- If you are making your own class sets of Numeral Cards, see p. 24.
- If you plan to provide folders in which students will save their work for the entire unit, prepare these for distribution during Session 1.

Session 1

Getting Started with Interlocking Cubes

Materials

- Sample object made of interlocking cubes
- Interlocking cubes (100+ per pair)
- Stick-on notes (for labeling objects)
- Student Sheet 1 (1 per student)
- Transparency of Student Sheet 1
- Chart paper (optional)
- Overhead projector

What Happens

Each student or pair builds an object with interlocking cubes. In small groups, students estimate how many cubes were used to make each object. Students count the cubes in their objects and then write about the activity. Their work focuses on:

- building with interlocking cubes
- estimating numbers of cubes
- counting cubes
- writing about math

Activity

Introducing the Mathematical Environment

Begin this session with a brief discussion about some of the tools you'll be using and the kind of work students will be doing in this unit. See the **Teacher Note,** Introducing Materials (p. 9), for hints about establishing routines for using and caring for manipulatives.

In our mathematics class this year, you will be using many different tools to help you solve problems and to show people how you are thinking about a problem. During the next few weeks, we will be using tools like calculators, pattern blocks, 100 charts, and interlocking cubes as we play math games and solve math problems.

❖ **Tip for the Linguistically Diverse Classroom** Ensure that this introduction is comprehensible to all students by showing the different mathematical tools as you mention them.

Ask students if they are familiar with any of these materials and how they have used them in the past.

Mathematicians use lots of different tools when they solve problems. Mathematicians also show us how they think about and solve problems. They talk about their work, draw pictures, build models, and write about their work so they can share their ideas with other people.

When you are working on a math problem, I will often ask you to use words, pictures, or numbers to explain how you solved it. Lots of times I will ask you to talk about how you solved a problem—either with a partner, in a small group, or with the whole class. This is one of the ways we can share good ideas and strategies for thinking about the math problems we are working on.

Many of you will invent ways of solving problems on your own, and I look forward to hearing all your ideas.

Activity

Building with Cubes

Pass out containers of interlocking cubes to groups of four to six students who will work together. Give students time to experiment with how to put the cubes together to make objects. Then tell them they are to build an object, on their own or with a partner, that will be used in the next activity.

Have the class brainstorm objects they can make (for example, chair, table, car, and so on). Allow 10 to 15 minutes for this, and give a 3-minute warning for students to complete their object and put away extra cubes. Students can label their objects with stick-on notes or small pieces of paper:

When your object is ready, put it in front of you so others can see it. Write the name of your object on a small label [*demonstrate*], and put the label in front of your object.

Give students a chance to look at each other's objects. You might allow them to walk around and view the objects, talking quietly, as if they were in a museum.

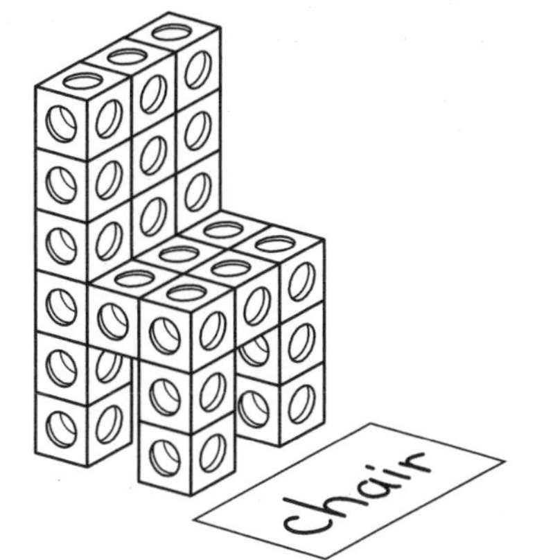

Activity

How Many in Each Object?

Display the transparency of Student Sheet 1, How Many Cubes in Each Object? on the overhead or draw the chart on the board. Fill in your name in the first row. Show students your sample object, and write its name on the chart. For example:

Name of maker	Name of object	Estimate	Count
Ms. Brown	Chair	15-30	26

Carry your object around to show to students. Ask them to estimate how many cubes you used.

Talk quietly in your groups and see if you can agree on an estimate of about how many cubes I used in my object. For example, you might say something like "around 30" or "about 15." If your group can't decide on one number, give a range as your estimate—maybe "between 35 and 40." Decide on an estimate that seems right to all the people in your group.

As groups are ready to share their estimates, list their ideas on the board. Ask a few students how they came up with their estimates. Support any use of systematic grouping, such as: "There are 4 legs and they each have 3 cubes, so that is 3, 6, 9, 12. Then the top part looked like about 10." On your chart, write the range of estimates for the number of cubes in your object, as shown above.

Making Estimates for Student Objects Hand out Student Sheet 1, How Many Cubes in Each Object? In the first row, each student writes your name, the name of your object, and his or her estimate as to how many cubes it took. Explain that their individual estimate may be different from their group's estimate. Whatever number they write down, they are not to change it after you count.

When everyone has written an estimate, disassemble your object into countable pieces (some single cubes, some in groups) and count the cubes. Count by groups when appropriate, rather than all by ones. Students write this actual count on their charts in the last column. They can compare the last two columns to see how good their estimates were.

Now you will estimate the number of cubes in each of the objects that the other people in your group made. Work with one object at a time. In your chart, write who made the object, the name of the object, and an estimate of the number of cubes in it.

❖ **Tip for the Linguistically Diverse Classroom** Students who have difficulty writing in English may draw simple pictures in the "Name of object" column.

Advise students not to pick up the objects as they make their estimates, as the objects might break apart.

Look closely, but don't count all the cubes. You are only trying to guess *about* how many. You may talk about your estimates with classmates, but you do not need to agree on them. Write down the number that *you* think is right.

When everyone in your group has written an estimate, the maker of the object will take it apart and count the cubes. Everyone in the group writes down that actual count.

Don't hurry. Work together. By estimating one object at a time and learning the actual number of cubes, you may make a more accurate estimate of the *next* object.

It's important not to change your estimates after you have written them down. Don't worry if they are not as close to the actual count as you wish. You will become better estimators with practice.

Get around to groups quickly to make sure they understand the task and to clear up any confusion. You might appoint one student in each group to help the others fill in their charts and to be sure everyone has written an estimate before counting. Ask some students to show or tell both you and their group how they are estimating.

Activity

Writing About the First Session

For the remainder of the session, students write in responses to the questions on Student Sheet 1, How Many Cubes in Each Object?

- Tell about what you enjoyed in math class today.
- Suppose someone was trying to estimate how many cubes were in an object. What advice would you give that person?

❖ **Tip for the Linguistically Diverse Classroom** Give students the option of responding to these questions in their native language. If they are not yet writing, they may communicate their answers orally to a family member, as homework. Ask for an adult signature at the bottom of Student Sheet 1, signifying that the child answered the questions.

Students may need some ideas about how to respond to these questions. Some fourth graders have said they enjoyed building, cooperating, estimating, and doing other things that weren't like real math.

To show that you value students' ideas about math, you might copy all their answers to the second question onto chart paper for display. Some answers that one fourth grade teacher recorded are given in the **Dialogue Box**, Advice About Estimating Numbers of Cubes (p. 10).

If you have prepared folders or math notebooks for students to keep their work in, this is a good time to pass them out. Students can put their names on the folders and put Student Sheet 1 inside.

Name Saloni Date Sept 10

Student Sheet 1

How Many Cubes in Each Object?

Name of maker	Name of object	Estimate	Count
Rebecca	dragon	33	35
Joey	Zap	50	49
Rafael	Gate	45-50	47
Shoshana	wink	25-30	23
Jesse	pencil holder	40	40

Tell about what you enjoyed in math class today.

I enjoyed estamating the Objects. and I lerned to estamate better.

Suppose someone was trying to estimate how many cubes were in an object. What advice would you give that person?

To look at them and then brake them apart and then estamate the peces and add them all together.

 99 *Investigation 1 • Session 1* *Mathematical Thinking at Grade 4*

Introducing Materials

Teacher Note

Using concrete materials in the classroom may be a new experience for many students and teachers. Before introducing new materials, think about how you want students to use and care for them and how they will be stored in your classroom.

Introducing a New Material Students will need time to explore a new material before using it in a structured activity. By freely exploring a material, students will discover many of its important characteristics. Once having had the chance to make their own decisions about how they would like to use a material, students will be more ready to use it in a focused way when it is introduced as a tool in a specific activity. Although some free exploration should be done during regular math time, many teachers make materials available to students during free time or before or after school.

Establishing Routines for Using Materials Establish clear expectations about how materials will be used and cared for. Students generally find math manipulatives attractive and engaging, and are eager to use them. Consider having students suggest rules for how materials should and should not be used. Students will often be more aware of rules and policies that they have helped create. You might want to establish the rule that if students are unable to use materials responsibly, they lose the opportunity to use them for a period of time.

Plan how materials will be distributed and cleaned up at the beginning and end of each class. Some teachers assign one or two students each week to be responsible for passing out and collecting materials. Most teachers find that stopping 5 minutes before the end of class gives students time to clean up materials, put their work in their folders, and double-check the floor for any stray materials.

Caring For and Storing Materials Store manipulatives where they are easily accessible to students. Many teachers store materials in plastic tubs or shoe boxes arranged on a bookshelf or along a windowsill. Students should view materials as useful tools for solving problems or illustrating their thinking—the more available they are and the more frequently they are used, the more likely students are to use them. Model this process by using manipulatives yourself when you are solving a problem, and encourage all students to use them.

If manipulatives are only used when someone is having difficulty, students can get the mistaken idea that using materials is a less-sophisticated way to solve a problem. Mathematicians frequently use concrete materials and build models to solve problems and to explain their thinking. Encourage your students to think and work like mathematicians!

D I A L O G U E B O X

Advice About Estimating Numbers of Cubes

Student Sheet 1 asks students to give advice on how to estimate numbers of cubes. There are no right or wrong answers to this question. In considering the matter, students think about estimation and how it differs from counting. One fourth grade teacher, who copied her students' advice onto chart paper and posted it on the wall, found that students gave advice of several different types.

Some students described what estimation is:

Kyle: Look at it [the object]. If the time to guess is over, try to think what is closest, and then guess what you think.

Nadim: I would tell another person not to just take a wild guess. Try to figure it out, but don't just count the cubes.

Rebecca: You have to look very hard and kind of count.

Luisa: Don't just guess. Look at it good and then guess.

Others gave advice about how the number must relate to the size of the object.

Nick: Try to guess on one that looks the same size.

Lina Li: Look at the size of the thing and try to guess.

Jesse: Make a big estimate for a big object and a small one for a small object.

Others gave more specific advice about how to count or compute without counting every cube.

Rashaida: Count some, and then times by that number.

Teresa: I count one [row] and I know there are 12 [*here she drew a 3 by 4 rectangle*].

Karen: Count one part, and then times it by as many times as you need.

Vanessa: You could count by 2's if it's hard for you to count by 1's.

DeShane: Count by a higher number than 1.

How Many Hundreds?

What Happens

Pairs count out 100 interlocking cubes and group them to show clearly that the total is 100. Groups of students then estimate how many hundreds there are in the box of cubes their group received. The teacher records the estimates from all of the groups and students figure out how many hundreds there are in the total of all the estimates. After counting their cubes, groups estimate how many hundreds of cubes there are in the class total. Finally, individuals figure out exactly how many cubes there are altogether.

- arranging cubes to show how many
- estimating how many hundreds of objects
- estimating hundreds in a sum of numbers
- keeping track while counting
- adding three-digit numbers
- using calculators

Materials

- Containers of interlocking cubes
- Student Sheet 2 (1 per student)
- Chart for How Many Cubes? transparency
- Student Sheet 3 (1 per student)
- Student Sheet 4 (1 per student, homework)
- Student Sheet 5 (1 per student, homework)
- One-centimeter graph paper (1–2 per student, homework)
- Calculators (at least 1 per pair)
- Family letter (1 per student)
- Chart paper (optional)

Ten-Minute Math: Estimation and Number Sense Two or three times during the next week, try the Ten-Minute Math activity Estimation and Number Sense. These activities are designed to be done in any 10 minutes outside of math class, perhaps before lunch or at the end of the day.

In this activity, students estimate an answer to a problem that you show for a brief time. Then they see the problem again and find a more precise solution. Begin with problems like those students are doing in the main activity in these sessions, How Many Hundreds? Present a problem on the chalkboard or overhead. For example:

$53 + 404 + 248 + 99$

How many hundreds are in all these numbers together?

Allow students to think about the problem for about a minute. Cover up the problem, and ask for students' estimates. Then uncover the problem and continue the discussion.

You might write the four numbers on four cards and ask students how they could rearrange the cards to make it easier to find how many hundreds (for example, 53 + 248 is approximately 300).

For a full description and variations on this activity, see pp. 91–92.

Activity

Counting Out 100 Cubes

Tell students that today they will be arranging cubes in hundreds so they can find out how many cubes the class has altogether.

Work with a partner to arrange exactly 100 cubes in a way that makes it easy for me to count them. I don't want to have to count the cubes by 1's; I want to be able to count them quickly. What would be a good number to count by—one that would help me count to 100 easily? (5's, 10's, 20's)

Distribute the interlocking cubes so that each group has a box or bucket. Pairs (or individuals, if there are enough cubes) each take 100 cubes and group them in some way to show that they have 100. Some pairs may decide they will each take 50 to arrange, and then put their 50's together.

You may need to help some students get started. Suggest they look around at what their classmates are doing and see if they can find an easier way to count the cubes. As students are working, observe each pair. Move quickly to students who appear not to understand the task. When it is unclear how they are making their 100, ask them questions like these:

How can I tell by counting just *part* of your arrangement how many cubes you have?

What will your 100 (or 50) cubes look like when you finish?

How many cubes have you arranged already?

Students who finish making their 100 early can help others in their group finish. Or, if there are enough cubes, they might count out another 100 and arrange it in a different way from their first 100.

Draw on the board or on chart paper some of the ways students arranged 100 cubes. Label any rectangular arrays with dimensions, as in this 5-by-20 array:

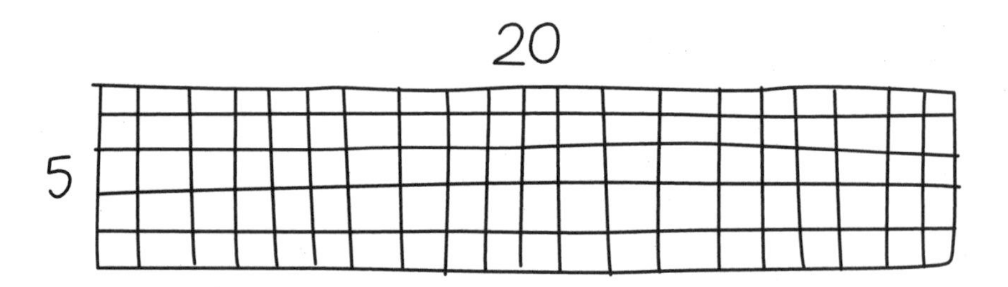

Call attention to the drawings you have made. Ask which students made each of these arrangements, and whether there are any arrangements you missed. Some students may make three-dimensional shapes. Add these to your drawings. Do not try to draw all shapes; those that are difficult to draw may also be difficult to count, and thus don't fit the task you posed. If students would like you to draw their irregular shape, enlist their help in explaining how you should make the drawing.

Writing About 100 Hand out Student Sheet 2, Making Hundreds, to each student. This sheet asks students to draw and write about their work with the cubes:

1. Make a drawing to show how you grouped 100 cubes.
2. How do you know you have exactly 100 cubes?
3. What other ways could you have arranged 100 cubes?

As they finish this writing and drawing task, students leave their hundreds grouping out near them, and group any extra cubes together on the table.

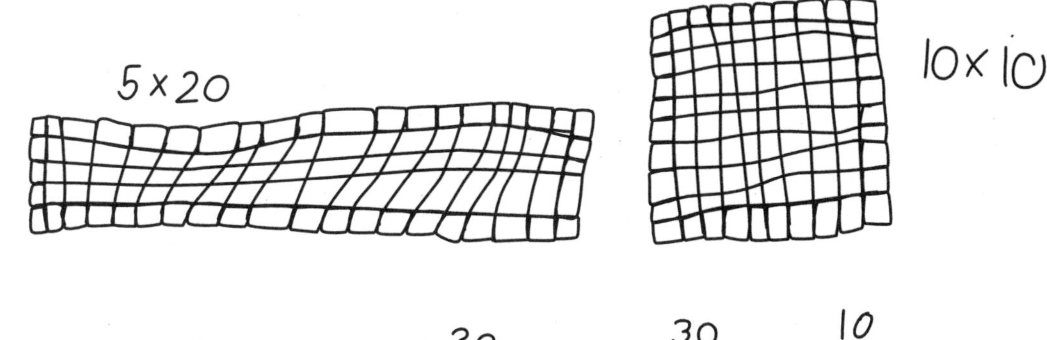

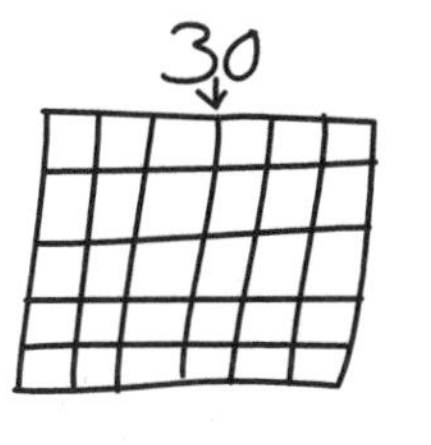

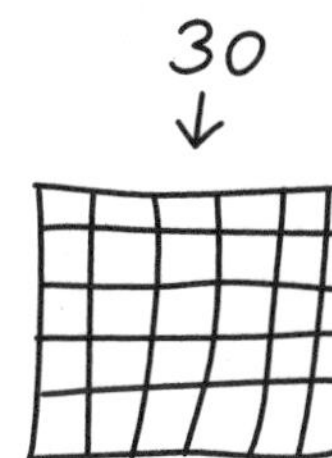

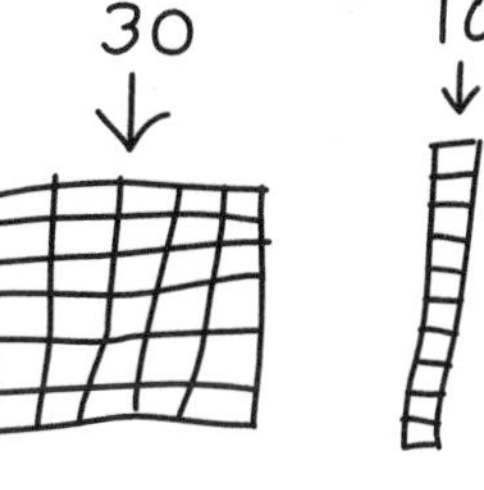

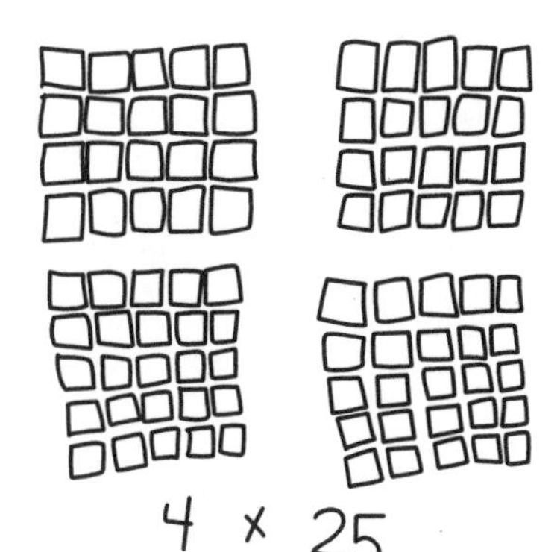

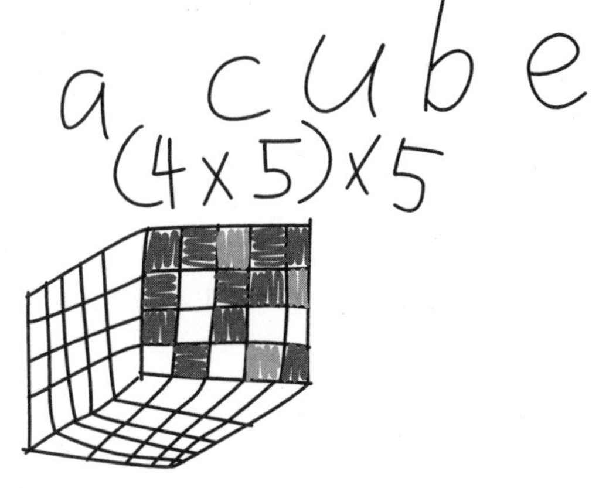

Activity

Estimating How Many Cubes Altogether

When students have finished writing about their hundreds, ask them to turn over their sheets. Pose this problem:

We're going to make an estimate of all the cubes we have in the whole class. But to start with, each group must come up with one number as an estimate of how many cubes *they* have.

How many cubes do you think were in your box when I handed them out? Think about the hundreds that people in your group have made, and think about the extras. Group the extras into piles that look to be about 100. Don't count them; just estimate groups of 100.

Everyone in your group must agree to one estimate of the number of cubes in your box.

As students are working, set up the Chart for How Many Cubes? transparency, or set up a chart like the following on the board or chart paper. After 3 or 4 minutes, write the groups' estimates on your chart. (The "counts" will be added in the next activity.)

How Many Cubes?		
Group	*Estimates*	*Counts*
Rebecca's group	350	
Qi Sun's group	315	
Kenyana's group	450	
B.J.'s group	300	
Marci's group	400	

Finding the total of all the estimates will be more challenging if you can get at least two estimates that have partial hundreds in them. If all of the groups give answers in exact hundreds, ask about the extras that didn't group into hundreds.

You said about 400—but would 350 or maybe 450 be a closer estimate than 400?

When your chart includes estimates for every group, ask students to figure out how many hundreds of cubes there are in the total of all the estimates *without writing anything.*

Talk quietly with your partner to figure out the number of hundreds in the total of all the groups' estimates. Figure this out in your head, without writing down the numbers.

Take a poll of the different answers, and tally how many pairs found each answer.

Fourth graders will often count only the hundreds in the hundreds column. They would, for example, get 1700 for the numbers 350, 315, 450, 300, and 400, adding 3 + 3 + 4 + 3 + 4. Ask one of these students to explain his or her thinking. Then ask a student who gets a higher answer (1800 in this case) to explain that answer. See the **Teacher Note,** Estimating Is Not Rounding (p. 19), for information on ways students may confuse rounding with good estimation strategies.

Note: At this level, some students, when hearing an answer like "seventeen hundred," will think that their answer of "one thousand, seven hundred" is different. Discuss these two ways of saying numbers that are larger than 1000. Students can count by hundreds to 1000 and find how many hundreds are in 1000.

Activity

Counting How Many Altogether

As in earlier activities, students now check their estimates by making actual counts. Each group counts all their cubes once, writes down the total, then counts again to check. They can split up the work in any way that makes sense to them and involves everyone in either counting or checking. For discussion later, notice the ways that groups organize the counting. Encourage them to keep the arrangements of hundreds they have and to make more arrangements of exactly 100 so they can easily check their counting.

Students record the total of each count on the back of Student Sheet 2. Collect the totals from the groups as they finish checking, and display them in the Count column on the How Many Cubes? chart you started in the preceding activity.

Ask representatives of groups to tell the class all the different ways they organized their counting. You might make a list, *Easy Ways to Count Many Things,* and post it for future reference.

Many teachers find this is a good place to end Session 2; students can go on to find the number of hundreds in the whole class in Session 3.

Note: Refer to the suggestions on p. 19 for suitable homework to follow Session 2.

Activity

How Many Hundreds?

Display again the How Many Cubes? chart you have been keeping on a transparency, chart paper, or the board. Pose the same question you asked earlier about the group estimates:

About how many hundreds are in all the numbers you counted? How many hundreds of cubes do you think are in all the boxes? Take your time to answer this question. Try to think of an easy way to do it—a way that you could teach someone else.

After you have an answer, talk with your neighbors about it. Change your answer only if they really convince you. If you do change your answer, explain how you got the new answer.

Write all the answers on the board. As before, some students are likely to count only the hundreds in the hundreds place. Others will combine the parts smaller than 100 to make more hundreds.

Ask two or three volunteers to explain how they got their answer. Many students are not accustomed to listening to other students. Use this as a chance for students to practice explaining their thinking and listening to each other.

Here is another chance for you to decide which is the best answer. Listen to how other people in class thought about this. Be ready to change your original answer or to defend your answer more firmly.

Hand out Student Sheet 3, How Many Cubes in the Class? Explain what they are to write in each column.

In the first column of this chart, copy down all the counts from the different groups.

In the second column, write how many hundreds you think are in all the counts. Then write how you figured it out in your head.

We'll do the third column later today.

Circulate to help students with this task. When students have trouble explaining how they got their answer, ask, "How do you know the number of hundreds?" If they can explain how they know, tell them to write that explanation down.

Name Sean Date 9/15

Student Sheet 3

How Many Cubes in the Class?

Number of cubes in each group	About how many hundreds altogether? Estimate.	Add the numbers to find the exact total.
509 470 417 538	1800 Show or explain how you got your answer. Each group has 400 and some extra. 4 groups is 400 x 4 = 1600 and I gessed 200 extra.	

Activity

Finding How Many in All

Once students have completed the first two columns of Student Sheet 3, they work alone to fill in the third column, finding the exact total of the cubes in the class.

What is the total number of the cubes we counted in our class? You can figure this out in any way you want. Write the answer in the third column, where it asks for the exact total. Show or explain how you figured it out.

As students are finishing, suggest that they compare their two answers.

Check to see how close your exact answer is to the answer you estimated. If your answers are very different, estimate again. See if you can find out what went wrong. Ask you partner to help you.

Observing the Students While students are working, you can learn a lot about their familiarity with numbers and addition by observing what they do. This entire unit is designed to help you assess your students' understanding of mathematical ideas, as illustrated by the sample assessment questions on p. I-20. Do not expect that all beginning fourth graders will be able to add so many large numbers, but use this chance to see what they *can* do. See the **Teacher Note,** Two Powerful Addition Strategies (p. 21), before evaluating their work. As you watch them, consider the following:

- Do students recognize the numbers they are adding? Does their method of addition take into account the values of the numbers?
- Do they work from right to left as in a traditional algorithm, or from left to right, adding larger values first? If they work from right to left, can they explain what they are doing? What they are "carrying"?
- Do they group numbers in a way that makes them easier to add?
- Can they interpret their answer and compare it to their estimate of the number of hundreds?
- Do they attempt repairs if their estimate and answer are very different?

Students who finish writing early might put away their group's cubes in stacks of ten. Let students know that from now on, they will store the cubes in stacks of ten to make counting and computing easier.

Adding with the Calculator When students are ready, distribute calculators for them to check the total number of cubes. If students have written the numbers under one another, they have probably written only one plus sign. Write the problem on the board horizontally (for example, 345 + 267 + 411 + 318 + 429 =), and remind students to put a plus sign in front of each number when adding on the calculator.

You can help students by dictating the numbers and pluses to them, or suggest that students do this for each other. Students can check their answers with their partners and with other students nearby.

If your students are not familiar with using calculators, allow some time for exploring. The **Teacher Note,** Introducing Calculators (p. 20), offers some ideas about how you might do this. At the end of class, put calculators in a place accessible to students, and explain that they are there for anyone to use when they need to. If you have enough calculators for each student to have one, you might let students keep them in their math folders.

Sessions 2 and 3 Follow-Up

Homework

A Design with 100 Squares Send home the family letter or the *Investigations* at Home booklet. After Session 2, give each student a copy of Student Sheet 4, A Design with 100 Squares, and one or two sheets of one-centimeter graph paper. At home, they make a design using exactly 100 squares. On Student Sheet 4, they write, "I know this is 100 squares because . . ."

How Many? After Session 3, send home Student Sheet 5, How Many Hundreds? How Many Altogether? The two problems are similar to the one on Student Sheet 3. Students estimate the answers to the nearest 100, then add the numbers to get an exact total.

Estimating Is Not Rounding

Teacher Note

Ways to round numbers, often taught as arbitrary rules out of context, have frequently confused students. One fourth grader demonstrated this use of badly remembered rules in estimating the price of four items at $3.49 each. By the rule of rounding down if less than half a dollar, he rounded the $3.49 down to $3.00, and then he rounded the $3.00 down to $2.00 "because 3 dollars is less than 5 dollars." He then multiplied 4 × $2.00, and concluded that 4 × $3.49 is about $8.00—nowhere near the $14.00 that he would have arrived at by doubling and redoubling three dollars and fifty cents (4 × 3.50), a result that is only 4 cents away from the exact answer.

Done sensibly, rounding is a useful skill. Consider these numbers of cubes:

423, 352, 475, 360

If we simply round each number to the nearest hundred and add them, we get 1700 cubes. However, if we look at the *value* of the numbers while rounding, we can pair up numbers to get a more accurate estimation. Adding the hundreds, we get 1400; then putting the 52 with the 60, we get just over another 100, making a little more than 1500. Finally, putting the 75 with the 23, we get just under another hundred, for a total of about 1600. (Actual total: 1610.)

Many times, the sum of rounded numbers makes a good estimate, and rounding is an efficient estimation strategy. However, teaching routine rounding rules discourages students from paying attention to the values of the numbers. Practicing sensible estimating, on the other hand, develops good number sense. Routine rounding rules focus students' attention on the digits furthest to the right, which make up the least part of the numbers. Good estimation strategies teach us to look first at the digits with highest value—those furthest to the left.

Teacher Note *Introducing Calculators*

If students have not used calculators during mathematics class before, they will at first be distracted by having them. As with any new material, students need time to explore the calculator and find out what it can do. As you circulate, make sure students know how to clear the calculator between problems and how to use the +, –, and = keys. If they are comfortable with the + and – keys, ask them if they can make up a problem using the × or ÷ keys to see if they get the answer they expect. Get a sense of how comfortable students are with the calculator:

- Can they do straightforward computation easily?
- Do they know how to recover if they make a mistake?
- Are they familiar with the symbols on the keyboard?
- Can they read the screen?

When students seem reasonably comfortable, you might talk about using the equals key as a repeat key:

Clear your calculator. Then press the plus key, the number 2, and the equals key. Watch your calculator display while you slowly press the equals key again, and then again. What is happening? [*The calculator is counting by 2's.*] **Stop for a moment. Guess what number the calculator is going to display next.**

Write on the board:

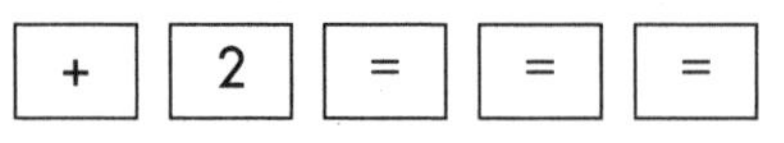

You might also write answers below the equal signs as students say them:

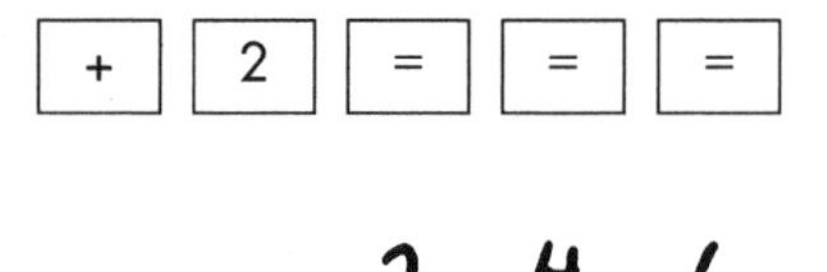

(Don't write the answers *after* the equal signs, as the statement would be untrue.)

Once students have tried the repeating equals, suggest that they try it with a different number.

Choose another number you would like the calculator to count by. Press the plus key, the number you want the calculator to count by, and then the equals key. Guess what the next number will be before you press the equals key. Then guess the next number and press equals again; guess and press again. Do this until you get over 100.

Two students sharing a calculator can present puzzles to each other. While one student looks away, the other presses +, a number, =, and then presses = a few more times. The other student must guess by pressing only = what number the calculator is counting by (that is, the number the first student chose).

If you use this puzzle to challenge students, choose an appropriate level of difficulty for each student. As puzzle numbers, start with 5, 10, 3, or 4; then, for students who figure these out easily, use 12 or 15.

This task is a good informal diagnostic. Ask yourself:

- Do they understand the task?
- Do they understand that to find out what the calculator is counting by, they must find the difference between two consecutive numbers? Do they see that to decide what number will appear next, they must add on the number being counted by?
- Can they add easily, or can they count on?
- Do they recognize that the sums of ones digits repeat? (For example, 9 + 3 is 1$\underline{2}$, so 9 + 63 is 7$\underline{2}$.)

If any students have difficulty with the calculator, you may need to work with individuals or schedule other sessions like this one. Rather than teaching the use of the calculator to the entire class at once, work with small groups. You might ask students who know how to use the calculator to work with those who don't.

Two Powerful Addition Strategies

Teacher Note

By fourth grade, many students know the traditional algorithms for addition and subtraction ("carrying" and "borrowing"), and some can apply them with ease when they encounter a clear-cut addition or subtraction problem. But many real mathematical problems do not shout out "I am *this* kind of problem! Use *this* operation to solve me!" Students have learned that when they have two numbers, they need to do an operation to get the right result. What they often don't understand is that a major piece of mathematics is figuring out what operation to use and why it works.

Even when students have done the hard work of determining whether their problems involve addition or subtraction, we want to discourage them from automatically applying a correct algorithm. Why? Again and again in classrooms, we have seen students incorrectly apply or incorrectly remember memorized algorithms. As soon as the "correct" algorithm is introduced, students seem to forget their own valuable strategies and understanding of number relationships in favor of a procedure they believe they are expected to use. For example, we have often seen students make the following type of error:

$$\begin{array}{r} {}^{1}17 \\ 17 \\ +\,17 \\ \hline 42 \end{array}$$

These students are saying to themselves the familiar chant, "put down the 2 and carry the 1." In their focus on the mechanics of this rule, they fail to see that they should instead "put down the 1 and carry the 2."

Most of us who are teaching today learned to add in the manner these students are attempting, starting with the ones, then the tens, then the hundreds, and so forth, moving from right to left and "carrying" from one column to another. While this algorithm is efficient once it is mastered, there are many other ways of adding that are just as efficient, are closer to the ways we naturally think about quantities, connect better with good estimation strategies, and generally result in fewer errors.

When students rely only on memorized rules and procedures they do not understand, they usually do not estimate or double-check. They make mistakes by rote that make no sense if they think about the numbers. We want students to use strategies that encourage, rather than discourage, them to think about the quantities they are using and what a result is likely to be. We want them to use their knowledge of the number system and important landmarks in that system. We want them to easily break apart and recombine numbers in ways that help them make computation more straightforward and, therefore, less prone to error. Writing addition and subtraction problems horizontally rather than vertically is one way to help students focus on the whole quantities. Varying the way you write problems will help students become more flexible in their addition and subtraction strategies.

The two powerful addition strategies discussed here are familiar to many competent users of mathematics. If you encourage students to come up with their own strategies, some of them will probably invent others. It is critical that every student have more than one way of adding so that an answer obtained using one method can be checked by using another. Anyone can make a mistake while doing routine computation—even with a calculator. What is critical, when accuracy matters, is that you have spent enough time estimating and double-checking to be able to rely on your result.

Left to Right Addition: Biggest Quantities First When students develop their own strategies for addition from an early age, they usually move from left to right, starting with the bigger parts of the quantities. For example, when adding 27 + 27, a student might say "20 and 20 is 40, then 7 and 7 is 14, so 40 plus 10 more is 50 and then 4 more makes 54." This strategy is both efficient and accurate. Some people who are extremely good at computation use this strategy as their basic approach to addition, even with large numbers.

One advantage of this approach for students is that when they work with the largest quantities first, it's easier to maintain a good sense of what the final sum should be. Another advantage is that students tend to continue seeing the two 27's as whole quantities, rather than breaking them up into their digits and losing track of the whole. Using the traditional algorithm ("7 and 7 is 14, put down the 4, carry the 1"), students too often see the two 7's, the 4, the 1, and the two 2's as individual digits. They lose their sense of the quantities involved, and if they end up with a nonsensical answer, they do not see it because they believe they "did it the right way."

Rounding to Nearby Landmarks Changing a number to a more familiar one that is easier to compute with is another strategy that students should develop. Multiples of 10 and multiples of 100 are especially useful landmarks for students at this age. For example, in order to add 199 and 149, you might think of the problem as 200 plus 150, find the total of 350, then subtract 2 to compensate for the 2 you added on at the beginning.

To add 27 and 27, as in the previous example, some students might think of the problem as 30 + 30, then subtract 3 and 3 to give them the final result. Of course there are other useful landmarks, too. Another student might think of this problem more easily as 25 + 25 + 2 + 2. There are no rules about which landmarks in the number system are best. It simply depends on whether using nearby landmarks helps you solve the problem.

If you have students who have already memorized the traditional right-to-left algorithm and believe that this is how they are "supposed" to do addition, you will have to work hard to instill some new values—that looking at the whole problem first and estimating the result is critical, that having more than one strategy is a necessary part of doing computation, and that using what you know about the numbers to simplify the problem leads to fewer errors. Students must develop procedures that make sense to them and will therefore be used with greater confidence and accuracy.

Close to 100

What Happens

Students play a game that involves arranging digits to make numbers that have a sum as near as possible to 100. Their work focuses on:

- finding pairs of numbers that equal 100 or near 100
- using place value
- adding and subtracting

Ten-Minute Math: Estimation and Number Sense Two or three times during the next week, continue to do the Estimation and Number Sense activity in any 10 minutes outside of math class. Devise problems in which looking at the whole problem first, and perhaps rearranging the order of the numbers, will make it easier to solve mentally. For example:

$3 - 2 + 6 + 7 - 6 =$

$13 + 28 + 7 + 2 =$

$5 + 300 + 3 + 22 =$

$$\begin{array}{r} 303 \\ 420 \\ +197 \\ \hline \end{array}$$

Students do not write or use a calculator while estimating the sums. Encourage all kinds of estimation statements and strategies. For students who have difficulty seeing how changing the order of numbers can make the problem simpler, write each number in the problem on a separate card so they can reorder them.

For a full description and variations on this activity, see pp. 91–92.

Materials

- Numeral Cards (1 deck per group)
- Close to 100 Score Sheet (2 per student; 1 for class, 1 for homework)
- How to Play Close to 100 (1 per student, homework)
- Numeral Card sheets (1 set per student, homework)
- Scissors for cutting out cards in class, as needed
- Student Sheet 6 (1 per student)

Activity

Introducing Close to 100

If you are making your own sets of Numeral Cards, cut apart the 44 cards for one complete deck. Enlist students' help to make many sets quickly. If you mark the back of each deck differently, the decks are easy to separate if they get mixed up in use.

In the game Close to 100, each player is dealt six Numeral Cards—each with a digit from 0 to 9 or designated as a wild card, which can be used as any digit. Players arrange four of the cards to make two numbers that have a sum as close as possible to 100. See the **Teacher Note,** Directions for Close to 100 (p. 28), for complete instructions and examples. To introduce the game, show the students a hand of six cards:

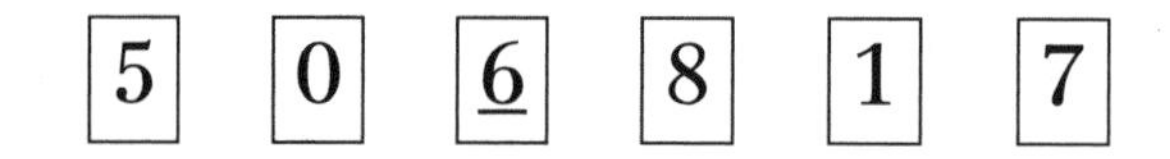

Ask them to suggest some numbers that could be made from two of these digits and write their suggestions down. Note that 08, which is not considered a two-digit number, is a legitimate answer. Then pose the problem:

We're going to play a game called Close to 100. Pick two of these numbers you suggested—or others that you can make—and add them together. Try to get a sum (a total) as close to 100 as possible. In fact, is there any way to get exactly 100 by adding two numbers made from these six cards?

As students are working individually on this problem, circulate and talk to as many as you can about strategies they are using. After a few minutes, collect a few solutions on the board. Explain the scoring of the game:

Your score for each round is how far your answer is from 100. Suppose you made 87 and 10: 87 + 10 = 97, and 97 is 3 away from 100, so your score for this answer would be 3. Suppose you made 60 and 51: 60 + 51 = 111. How far away is that from 100? (11) Your goal is to get the *lowest* score, so the score of 3 is better than this score of 11.

Display another set of digits for students to work with:

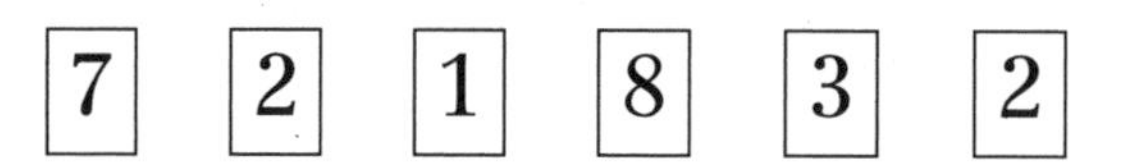

With this set, there are several ways to make exactly 100 (83 + 17, 87 + 13, 28 + 72, 22 + 78). Once you think students have the idea of making two numbers that have a sum near 100, they are ready to play the game.

To give students further practice with game strategy, before they play in pairs or small groups, play a game against the class. Deal out cards both for yourself and for the class. Students discuss ways to get close to 100 with their set of cards and why they think their answer is the closest possible to 100. Some students have difficulty articulating their thoughts. Posing some of the following questions may help them describe their strategy:

Is this the closest we can get to 100?
How do you know?

If students are unable to verbalize their thinking, keep the first two-digit numeral on the overhead, remove the other two cards, and substitute a different two-digit numeral.

Is this closer to 100 than the previous sum?

During your turn, you may want to describe your thinking process to the class as you consider your options.

Activity

Playing Close to 100

Students arrange themselves in pairs or groups of three. Each group needs one deck of Numeral Cards, and each student needs a copy of the Close to 100 Score Sheet. Explain the game:

To play Close to 100, deal each player six cards. Each player uses four cards to make two numbers with a sum as near 100 as possible. Players should check each others' combinations.

Write your numbers on your score sheet, then compute and write your scores. Discard the four used cards. You will then be dealt four new cards so you still have six cards for the next round. You can see on the score sheet that five rounds make a game.

Note: Students sometimes limit themselves by using multiples of ten whenever possible. To avoid this, you might leave out the Wild Cards when students first play, or establish a rule that Wild Cards can't be made into zeros.

Many students will play Close to 100 cooperatively, helping one another find the best combinations. Encourage this cooperation:

After you have dealt six cards to each player, take turns and help each other. Work together to get the best answer you can for each person.

Move among the groups to see that they all understand the game. Observe the strategies they use. If students seem stuck, make one two-digit number and ask them to make another number that will bring the sum close to 100.

Use this time to observe a few students whose thinking you would like to know more about. See the **Teacher Note,** Evaluating Skills During Games (p. 30).

While some students will soon invent strategies, others will have little idea how to think about the sum of two numbers without writing them one above the other. They will play Close to 100 by trial and error, trying many combinations and then choosing the one with the sum closest to 100.

Let students play until they understand the game well enough to teach it to someone else. Then find a time for them to prepare their own decks for playing Close to 100 at home. Students will need scissors and sets of Numeral Card sheets. If you think students have scissors at home and can do the cutting themselves, send the Numeral Cards sheets home. Class sets of Numeral Cards will be needed in Investigation 2, when this game will be a choice among several activities.

Save 20 minutes of Session 4 for the students to complete the Teacher Checkpoint.

Activity

Teacher Checkpoint

Thinking About Close to 100

Hand out Student Sheet 6, Problems for Close to 100, for students to do individually. Students may want to use Numeral Cards so they can move the numbers around to try different combinations. They can share the sets they have been playing with.

Circulate around the class to be sure students understand what is expected and have a way to begin. Observe their strategies. You may need to ask questions of individuals to understand their thinking.

- Are they able to add or estimate mentally, or do they need to write down each problem?
- Do they use a strategy, or do they try combinations almost randomly?
- If they begin randomly, do they narrow down the choices to reasonable numbers, or do they try combinations indiscriminately, adding pairs of two-digit numbers until they find a sum close to 100?
- Do students consider the sum of the tens digits when picking two numbers? For example, do they recognize that if they have a number in the twenties, they will need another number in the seventies or the eighties to be close to 100?

Be sure to save the class decks of Numeral Cards you used in this investigation for use in games throughout this unit.

Session 4 Follow-Up

Homework

Playing Close to 100 Send home copies of How to Play Close to 100 and a new Close to 100 Score Sheet with each student. Students also take home the Numeral Cards they have prepared (or a set of Numeral Cards sheets to cut out). Students play the game with family members, and report the next day on their experiences playing at home.

Advise students to find a safe spot at home to keep their Numeral Cards, because they'll be learning more games in the next couple of weeks to play at home with their cards.

Teacher Note

Directions for Close to 100

Close to 100 can be played as a solitaire game, but in class, two or three students will play together. Each group will need one deck of Numeral Cards and a Close to 100 Score Sheet for each player.

How to Play

1. For the first round, deal out six cards to each player.
2. Each player uses any four of these cards to make two numbers that, when added, come as close as possible to a total of 100. (See the sample round below.) Wild Cards can be used for any numeral.
3. The player records these two numbers and the total on the Close to 100 Score Sheet. The player's score for each round is the difference of the sum of the two numbers from 100. The four cards used are then placed in a discard pile.
4. For each successive round, four new cards are dealt to each player, so that all players again have six cards.

The game ends after five rounds. If the deck runs out of cards before the game is over, shuffle the discard pile and continue to deal. At the end of five rounds, players total their scores. The lowest score wins.

Sample Game

Round 1

Joey is dealt these cards:

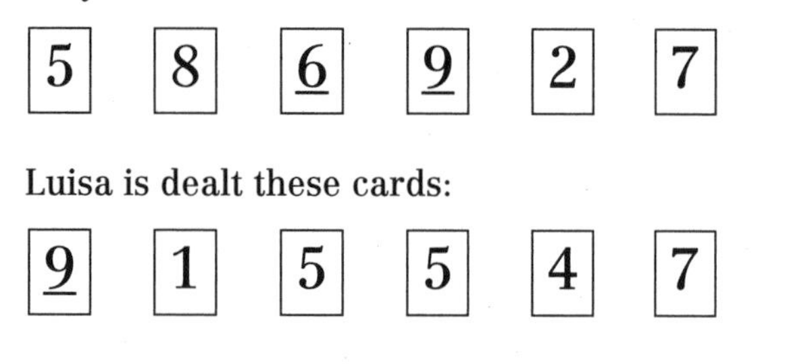

Joey makes 58 + 29, and Luisa makes 45 + 57.

Round 2

Joey has 6 and 7 left from round 1, and is dealt

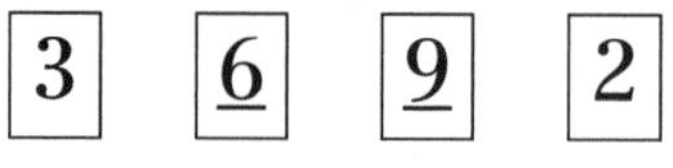

Luisa has 9 and 1 left from round 1, and is dealt

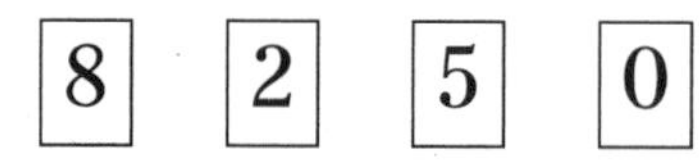

Joey makes 36 + 62, and Luisa makes 98 + 02.

Note: Both Joey and Luisa could have gotten closer to 100 in round 1, and Joey could have gotten closer to 100 in round 2. Can you see how?

The game proceeds, and their final scores look like this:

CLOSE TO 100 SCORE SHEET

Name Joey

GAME 1					Score
Round 1:	5 8	+	2 9	= 87	13
Round 2:	3 6	+	6 2	= 98	2
Round 3:	9 3	+	0 6	= 99	1
Round 4:	7 0	+	3 0	= 100	0
Round 5:	8 7	+	1 1	= 98	2
				TOTAL SCORE	18

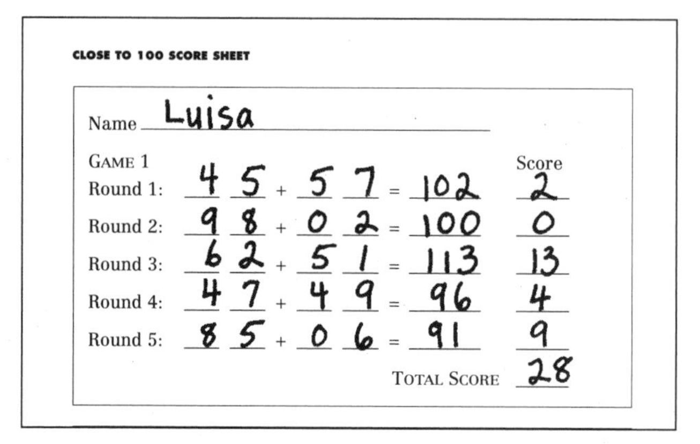
CLOSE TO 100 SCORE SHEET

Name Luisa

GAME 1					Score
Round 1:	4 5	+	5 7	= 102	2
Round 2:	9 8	+	0 2	= 100	0
Round 3:	6 2	+	5 1	= 113	13
Round 4:	4 7	+	4 9	= 96	4
Round 5:	8 5	+	0 6	= 91	9
				TOTAL SCORE	28

Joey has the lowest score, so he wins.

Scoring Variation: Negative and Positive Integers

Students should be very comfortable with the basic game before trying this variation. Its use is specifically suggested during Choice Time in Investigation 3.

In this variation, the game is scored with negative and positive integers. If a player's total is above 100, the score is recorded as positive. If the total is below 100, the score is negative. For example, a total of 103 is scored as +3 (3 above 100) while a total of 98 is scored as –2 (2 below 100). If using this variation, Joey's and Luisa's score sheets from the sample game would look like this:

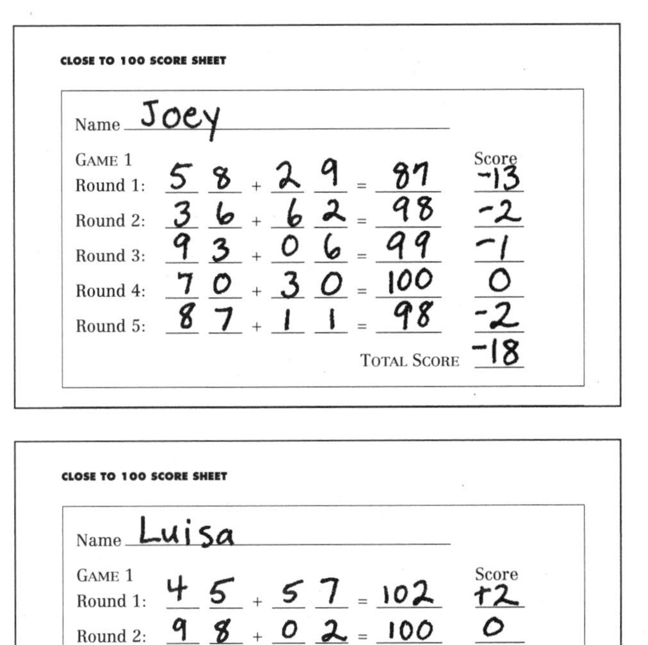

CLOSE TO 100 SCORE SHEET

Name Joey

GAME 1

				Score
Round 1:	5 8	+ 2 9	= 87	-13
Round 2:	3 6	+ 6 2	= 98	-2
Round 3:	9 3	+ 0 6	= 99	-1
Round 4:	7 0	+ 3 0	= 100	0
Round 5:	8 7	+ 1 1	= 98	-2
			TOTAL SCORE	-18

CLOSE TO 100 SCORE SHEET

Name Luisa

GAME 1

				Score
Round 1:	4 5	+ 5 7	= 102	+2
Round 2:	9 8	+ 0 2	= 100	0
Round 3:	6 2	+ 5 1	= 113	+13
Round 4:	4 7	+ 4 9	= 96	-4
Round 5:	8 5	+ 0 6	= 91	-9
			TOTAL SCORE	+2

The player with the total score closest to zero wins. So, in this case, Luisa wins (her +2 is 2 away from 0, and Joey's –18 is 18 away from zero).

Scoring this way changes the strategy for the game. Even though Joey got many scores very close to 100, he did not compensate for his negative values with some positive ones. Luisa had totals further away from 100, but she balanced off her negative and positive scores more evenly to come out with a total score closer to zero.

Teacher Note — Evaluating Skills During Games

Close to 100 is the first of several games students will be playing in this unit. We include games for several reasons. Students enjoy them, so they play them again and again and get lots of practice. Because numbers are generated randomly, responsibility shifts from the teacher to the students. Whether playing collaboratively or as opponents in a game, students are likely to correct each other, which takes from the teacher the onerous task of telling students they have made a mistake. Students will also learn strategies from one another.

Because you are freed from teaching and checking work, you have an opportunity to observe. You can spend time with students whose mathematical thinking you would like to know more about. Ask questions about students' strategies. Play collaboratively with a student if that feels comfortable. Stop a group, or possibly the whole class, to share an interesting hand that you observed students grappling with. Collect ideas and strategies to share with the class. Your students have much to teach you about difficulties and strategies.

Strategies that work in mathematical games involve thinking about the mathematics involved. In Close to 100, one good strategy is to try to make 9 (90) with the tens digits and 10 with the ones digits. Students who think this way are making use of the concept of place value. For some examples of students' strategies, see the **Dialogue Box,** Strategies for Close to 100 (p. 31).

Some students may not be accustomed to thinking about strategies. Instead, they might pick digits randomly to make numbers and add them with paper and pencil as they have been taught. When you talk with these students, encourage them to estimate mentally before they do written addition.

DIALOGUE BOX

Strategies for Close to 100

During this discussion of the strategies that students were using for making numbers in the Close to 100 game, only a few students spoke up. The teacher posted the students' ideas on chart paper, planning to add to them another day.

Think about how you chose numbers to try. Do you have any advice for your friends on how to get close to 100? Tell me something that *you* look for when you try to get close to 100.

Rikki: I first try to find two numbers that equal 10, like 7 and 3, or 6 and 4.

For the tens digits?

Rikki: Yes.

Alex: I'd look for 9, and put another number with it, and it might come close to 100.

When you started with 9, was it on the tens side or the ones side?

Alex: On the tens side.

So you started with 90? What else did you look at?

Alex: I tried to get 10 on the ones side.

Emilio: I tried to find a high number and then a little one, like ninety-something, and then if I had a zero I made something like zero five (05).

So you tried to make something in the nineties and a single-digit number.

Teresa: I tried to find a way to get 9, and then I tried to find a way to get 10.

So you looked for two numbers that would make 90 and two that would make 10?

Teresa: Yeah, I make 9 with the first two numbers [meaning the tens digits; for example, 70 + 20, or 80 + 10], and 10 with the second two. I don't want to make 10 with the first numbers [70 + 30], because then I'd need two zeros for the last ones—and you hardly ever have two zeros. So I try to get numbers that make 9 instead of 10, and then I try to make 10 in the ones place.

Most of you had a strategy, a plan for how to put the numbers down. You weren't just trying things. Do you have any other thoughts? Think about what you were doing.

Alex: My way that I said is like Teresa's. Or if I couldn't make a 9, I tried for an 8 with the first numbers so I worked with 80 and 20.

Teresa: You can't get 100 that way 'cause you can't make 10 and 10.

Alex: But I can get almost there with 9's or 8's.

INVESTIGATION 2

How Many Dollars?

What Happens

Session 1 and 2: How Much Money? Each pair sorts a package of play money and counts how much money they have, keeping track on paper as they count. Then pairs report to the whole class how they organized their counting, before exchanging bags of money to count again. They explore different ways to count coins—mentally, with a calculator, or paper and pencil—to discover a "best" way. Finally, students play a game that involves finding two groups of coins that total one dollar.

Sessions 3 and 4: Number Sense and Coins Students learn a new game, picking out amounts of money in real coins by feel—without looking. Then they are introduced to Choice Time, in which they choose from among three games—the new one, and games introduced in the previous sessions—while keeping track of how they spend their time.

Mathematical Emphasis

- Grouping coins for more efficient counting
- Recognizing values of U.S. coins
- Recognizing the decimal point on the calculator

What to Plan Ahead of Time

Materials

- Play money—paper $1 and $5 bills, and plastic pennies, nickels, dimes, quarters, and a few half dollars (Sessions 1–4)
- Resealable plastic bags for play money: 1 per pair
- Scissors (Sessions 1–2)
- Calculators (Sessions 1–2)
- Paper bags (bags must be opaque) of real coins, holding 2 quarters, 3 dimes, 3 nickels, and 5 pennies to make exactly $1: several for Choice Time (Sessions 3–4)
- Numeral Cards: all available decks for Choice Time (Sessions 3–4)

Other Preparation

- Duplicate student sheets and teaching resources (located at the end of this unit) in the following quantities. If you have Student Activity Booklets, copy only the item marked with an asterisk.

  For Sessions 1–2

  Student Sheet 7, Ways to Count Money (p. 108): 1 per student

  Coin Cards (p. 109): 1 set per 2–3 students for class work, and 1 set per student (homework)

  How to Play Collecting Dollars (p. 113): 1 per student (homework)

  Coin Value Strips* (p. 139): copy and cut apart to provide each student with a desk-top strip

  For Sessions 3–4

  How to Play Hidden Coins (p. 114): 1 per student (homework)

  Close to 100 Score Sheet (p. 135): 1 per student

  Choice List (p. 140): 1 per student (optional)

- Divide your play money to make a bag for every pair. Rather than counting out, distribute the money by random handfuls so there are similar, but not equal, amounts in each bag. Ideally, there would be at least $5 per bag, but less will work if you don't have enough coins. Label bags with numbers or letters to identify them.
- Decide on your rules for Choice Time and how students will record what activities they have done. See the **Teacher Note, About Choice Time** (p. 44), for guidance.

Sessions 1 and 2

How Much Money?

Materials

- Bags of play money (1 per pair)
- Coin Value Strips (1 per student)
- Student Sheet 7 (1 per student)
- Coin Cards sheets (1 set per 2–3 students, and 1 set per student, homework)
- Scissors
- How to Play Collecting Dollars (1 per student, homework)
- Calculators (available)

What Happens

Each pair sorts a package of play money and counts how much money they have, keeping track on paper as they count. Then pairs report to the whole class how they organized their counting, before exchanging bags of money to count again. They explore different ways to count coins—mentally, with a calculator, or paper and pencil—to discover a "best" way. Finally, students play a game that involves finding two groups of coins that total one dollar. Their work focuses on:

- recognizing the value of U.S. coins
- organizing and keeping track of counting money
- finding amounts in coins with a total of one or more dollars

Ten-Minute Math: Estimation and Number Sense As you proceed through Investigation 2, continue to do the Estimation and Number Sense activity in any 10 minutes outside of math class. Present problems related to estimating amounts of money. For example:

25¢ + 52¢ + 79¢ 8 × 39¢ $10.00 – $5.75

Some classes have enjoyed estimating totals on cash register receipts.

Milk	2.39
Carrots	.79
Chicken	4.43
Rice	2.69

Suppose I'm in the grocery store and I have only ten dollars. Do I have enough money to pay for these items? How many dollars do I need?

Students may bring in short receipts to challenge the class to find how many dollars are needed.

Students do not write or use a calculator while estimating the answers to problems that you display for about a minute. Encourage all kinds of estimation statements and strategies.

For a full description and variations on this activity, see pp. 91–92.

Activity

Counting Money

Hand out a bag of play money to each pair. Distribute the Coin Value Strips for students to tape to their desks for reference. Students work together in pairs, counting the play money in their bag. They will need paper to keep track of their counting. Point out that you are not providing a special sheet for recording their work, but that they will need to keep track carefully enough so they can explain their counting to themselves and to others.

As you observe them working, pick two or three pairs to report how they are organizing their counting. Stop the class briefly for these reports. Caution students not to say their total, because another group will be counting the same bag.

When students are finished, they write the total value on their paper, with the letter or number that identifies the bag they counted. Pairs then trade bags of money with another pair who is finished.

Students may be tempted to hurry to count one bag after another; discourage this. If they have extra time while waiting for another pair to trade with, ask them to record their counting in more detail or to check their total by counting in a different way.

To conclude this activity, pairs who exchanged bags get together to compare results. Whether they agree or differ about the amounts, they should show each other how they kept track of their counting. They may work together to count the money again if they want.

Activity

Adding Coin Values

Student Sheet 7, Ways to Count Money, lists the contents of three handfuls of coins. Students are asked to find the total value of each. Many fourth grade students, when they see such problems on paper, don't realize that they can add coin values mentally. Instead, they write the values and add them with paper and pencil. This activity is designed to encourage them to use the more efficient and practical method of adding mentally to count coins.

Demonstrate to the whole class the process of adding coins mentally. List several coins on the board:

1 quarter
2 dimes
1 nickel
2 pennies

How much money do you think this is altogether? We're going to count it mentally. [*Point to "1 quarter."*] **How much is the quarter worth? Whisper the value.** [*Pause for the answer. Write $ .25 on the chalkboard. Then point to "2 dimes."*] **Add the dimes to the quarter. How much is 25 cents plus 10 cents?** [*Pause.*] **Plus another 10 cents?** [*Pause. When students agree, put a line through the $ .25 and write $ .45.*]

Continue adding one kind of coin at a time, with students giving the accumulated total and you writing it in decimal form (this is to prepare for calculator work). Then do it again, this time counting the same coins in a different order: the nickel, quarter, pennies, and then the dimes. Doing the nickel first will raise the issue of how to write 5 cents as a decimal.

How can I write 5 cents using a decimal point?

If students suggest "point 5," explain that cents are always written with two numbers, and write $ .50 on the board.

But that is the way to write 50 cents. How could we write 5 cents?

Establish that amounts below 10 cents, written with a decimal point, have a zero between the point and the number.

After you have counted the coins on the board in a different order, ask students which order seemed easiest. Tell them that many people always start with the coins with the highest values, and add each lower amount in order. Other people group coins to make easy numbers to work with—such as 50 cents or 1 dollar. So, for this group of coins, they might group the two dimes and a nickel to make 25 cents, which combines with the other quarter to make 50 cents. Then all they have to add is the two pennies.

Counting Money in Other Ways Invite students to try the same problem in another way. Have play money and calculators available.

How could we find the total value of these coins in another way? What tools might we use? I'd like you to work in pairs to do this problem another way. You may use the plastic coins, or a calculator, or paper and pencil.

As students work, move quickly to any pairs you think might have difficulty.

Fourth graders who choose the calculator are likely to find this problem difficult if they use decimal points. The problem requires pressing 24 individual keys: .25 + .10 + .10 + .05 + .01 + .01 = . If students want to try this, suggest that one student call out what keys to press while the other does the entering. Some will discover that it's much easier to treat all the values as whole numbers of cents, and easier yet to group values, perhaps entering only 25 + 20 + 7 = . Let students struggle a bit to discover that using a calculator efficiently takes some planning.

As you circulate, you might stop from time to time to have pairs share with the class the way they are computing the total. Look, for example, for pairs who write the coin values in a vertical column and add them; for students who use play coins and add mentally as they touch each coin; and for pairs who are using the calculator in an efficient way.

Deciding on the "Best" Way When pairs have all completed the problem in another way, hand out Student Sheet 7, Ways to Count Money, one to each student. Emphasize that they are to find the total value of each group of coins *in more than one way*. Explain the reason for this:

When you've done these three problems, I want you to decide which is the best way to count coins. And to make a good decision, you need to try several different ways.

Pairs of students may consult with one another to compare answers and methods, but strongly discourage erasing and writing the answer another student got. Insist that students who have differing answers take time to show each other, step by step, how they did the problem. Together they can agree on an answer.

As you observe students working, ask them to explain what they are doing. Encourage those who write out every problem first to try some mental addition, perhaps writing only partial totals to help them keep track.

❖ **Tip for the Linguistically Diverse Classroom** Refer students to their desktop Coin Value Strips for help in identifying the coins listed on Student Sheet 7. Students may answer the advice question orally.

Follow up with a brief discussion of students' "advice to a friend" on what they feel is the best way to count coins. Ask them to give reasons for their choice.

Activity

Teacher Checkpoint

The Collecting Dollars Game

Becoming Familiar with the Coin Cards In preparation for playing the Collecting Dollars game, distribute scissors and a set of Coin Card sheets to each pair or group of three who will play together.

First cut apart the 32 Coin Cards. When this is done, work with your partners to make pairs of cards that add to one dollar.

This pairing task may be quite challenging at first. Make available the play coins for students to use as they need to.

Once students have made 16 pairs, they may shuffle the cards and try again. As there are two cards with each quantity, there is more than one way to make each pair.

If students have difficulty recognizing the coins, suggest that they write lightly (so it can be erased later) the value of each coin near its picture—25¢ near the quarters, 10¢ near the dimes, and so forth.

While students are sorting the Coin Cards into pairs that make one dollar, and later when they are playing the game, observe and ask questions:

- Can students recognize the coins and their values?
- Can students count by 5's, 10's, and 25's to determine the totals?
- Can students determine how much more money is needed to make one dollar?
- Can students recognize groupings of three, four, and five coins without counting (for example, do they know that five pennies is 5 cents without counting each penny)?

Introducing the Game As students seem ready, teach them how to play Collecting Dollars.

Players deal out eight Coin Cards, face up in the center, and place the remaining pack face down. They then take turns making pairs that equal one dollar. At the end of a turn, the player turns up new cards from the face down pack to replace any pairs made, so there are always eight cards available at the beginning of the next player's turn.

If a player finds that no pair among the eight cards makes one dollar, and all players agree, the player may shuffle all eight back into the pack, deal out eight new cards, and take another turn. The game ends when all the cards have been paired.

Students can continue playing until the end of the class session. They may prefer to play cooperatively, finding pairs but not keeping track of turns or of who found each pair.

Making Coin Card Decks for Use at Home If there is a possibility that your students may not have scissors at home, make time for them to cut out their own sets of Coin Cards. Don't let them take home the Coin Cards they made for class use, as students will need them again for the Choice Time activities in Sessions 3 and 4.

Sessions 1 and 2 Follow-Up

Homework

Playing Collecting Dollars Hand out copies of How to Play Collecting Dollars to send home with sets of Coin Card sheets (or cards prepared at school). Students can play the game at home, by themselves or with a family member. Tell them to practice so they will be better at finding pairs when they play again in school.

Sessions 3 and 4

Materials

- Choice List (1 per student, optional)
- $1 of real coins in opaque paper bags (several)
- Numeral Cards (all class sets)
- Close to 100 Score Sheet (1 per student)
- Coin Cards (from previous session)
- Bags of play money
- How to Play Hidden Coins (1 per student, homework)

Number Sense and Coins

What Happens

Students learn a new game, picking out amounts of money in real coins by feel—without looking. Then they are introduced to Choice Time, in which they choose from among three games—the new one, and games introduced in the previous sessions—while keeping track of how they spend their time. Their work focuses on:

- organizing themselves to play games
- making numbers whose sum is close to 100
- adding coins to make dollars

Activity

The Hidden Coins Game

Spend some class time introducing the Hidden Coins game that will be one of the options during Choice Time throughout Sessions 3 and 4. Explain and demonstrate the game as follows:

Empty the real coins from one of the prepared bags onto a table or the overhead, and count the money with the students ($1.00). Use the mental counting strategies they worked with earlier.

Return the coins to the bag. Ask a volunteer to name any amount of money less than one dollar. Suggest that you'd like to start with an easy task, perhaps an amount that can be made with only two coins. Then, talking about what you are doing but not looking into the bag, take the coins for that amount from the bag. Ask students after each choice whether the coin will help you reach the target. If you select a coin that you cannot use (for example, one that makes your total exceed the target), put it back in the bag. Ask students to double-check that your selected coins total the amount named.

Suggest that when students play together, they might begin by trying to remove just one particular coin at a time. When they are able to do this easily, they can start naming amounts that require two, three, or more coins.

Since this is real money, set up a structure with the students to be sure a full dollar is returned to each bag after each turn, and that all of the bags are returned to you at the end of class.

Activity

Choice Time: Playing Math Games

Introducing Choice Time The remainder of this session and the next will be Choice Time, and students choose from the new game and two estimation and addition games they have played in earlier sessions. These activities are designed to give students more experience with number sense and coins.

Choice Time is a format that recurs throughout the *Investigations* units, so this is a good time for students to become familiar with its structure. The **Teacher Note,** About Choice Time (p. 44), will give you ideas about how to set up and organize the choices, and how students might use a Choice List to keep track of their work.

Introduce your rules and students' responsibilities for Choice Time. Students should try each choice sometime during the two sessions. You might encourage them to play the games with several different classmates.

How to Set Up the Choices If you set up the choices at centers, show students what they will find at each center. If students will be playing the games at their desks, make sure they know where they will find the materials they need.

Choice 1: Hidden Coins—paper bags, each holding $1 in real coins (2 quarters, 3 dimes, 3 nickels, 5 pennies)

Choice 2: Close to 100—decks of Numeral Cards, copies of Close to 100 Score Sheet

Choice 3: Collecting Dollars—decks of Coin Cards, plastic coins (for those who need them)

Introduce the Three Choices Briefly remind students how to play the games they learned earlier. Close to 100 and Collecting Dollars have variations you can suggest to add interest. Rather than presenting these to the whole group, teach the variations to small groups as you observe they are ready for a different challenge.

Once students have decided on their first choice, they may begin. Some may need your help in finding a partner or in deciding where to start.

Choice 1: Hidden Coins

Students play this game in pairs, and take turns trying to pick coins out of the bag by feel, without looking. If you have a limited number of bags, be sure students take turns so that everyone gets a chance to play the game. Players begin by picking one coin at a time. When they are comfortable identifying single coins, they move on to amounts of money made with two or more coins.

Choice 2: Close to 100

Students play this game in pairs or threes. Players are dealt six cards each and work simultaneously, each with their own six cards, to make two two-digit numbers (or a one-digit number with a zero, such as 07) that have a sum as close as possible to 100. Their score for that round is the distance of the sum from 100. Scores from five rounds are added, and the player with the lowest total is the winner.

Scoring Variation: Negative and Positive Integers As appropriate, introduce the scoring variation described in the **Teacher Note,** Directions for Close to 100 (p. 28). In this version, a player's score is positive if the sum is over 100, negative if the sum is under 100. Thus, 103 scores 3; 97 scores –3. The new object is to get a score as close to zero as possible. These scoring rules change players' strategy—now they must try to balance scores over 100 with scores under 100. Don't expect students to understand right away how the scoring affects strategy. Tell them to play a practice game so they can see how the scoring works.

Introduce the new scoring to six or eight students at a time. Students who understand the idea of positive and negative changes can use the calculator to help them total their score. Once you have introduced the variation, students can split into pairs or groups of three to play it. Students who have learned the new version can teach other students as they are ready.

Choice 3: Collecting Dollars

The object of Collecting Dollars is to find pairs of Coin Cards showing coins with a total value of one dollar. Students may simply work independently or with a partner to find all the pairs in the deck, or they may play a game; see p. 39 to review the rules. If students have marked the coins on the cards with amounts, they can challenge themselves now by erasing the amounts and determining values from the coin images alone.

Variation: Whole Dollars If students become so accustomed to the 32 Coin Cards that they memorize the pairs, you can introduce an alternative form of the game: Players make combinations of two or more cards that equal *any* whole number of dollars. For example, a player could take three cards with 50¢, 70¢, and 80¢ for a total of $2.00. Players may decide to score either by number of dollars collected or by number of cards collected.

Observing the Students

As students work, plan to meet with pairs or small groups who need help on the activities, or with those you have not had much chance to observe. Encourage students to get help from other groups while you are busy. As needed, enlist the help of students who understand specific games to act as resources for their peers who have questions.

Once you are settled into playing, I will be visiting with different groups to learn about your thinking. I might teach some new ways to play the games. If you have a question or need some help, ask another group who knows that game for help or for permission to quietly watch them play.

As you circulate among the groups and observe students, you might look for the following:

- How are students making decisions about how to organize their time and activity?
- Are there too many (or not enough) activities going on at once?
- Are students keeping track of the choices they have completed?
- Are students able to handle the mathematics well enough to develop strategies in games?
- What strategies are students using?
- Which game gives students the most difficulty? What mathematics in that game needs more discussion and practice?

Save 5 or 10 minutes at the end of Session 3 to have a discussion about the games. Ask students to share what they have particularly enjoyed, and what has challenged them.

As a whole class, do some planning for the Session 4 Choice Time. Would students like help in learning how to play certain games better? Who will volunteer to teach or to play with students who are less familiar with a game or with a variation? See the **Teacher Note,** Collaborating with the Authors (p. 45), for more about your role and your students' roles in finding the right level of challenge for each of them.

You may decide to make adjustments in the number of choices offered in Session 4, to pair students who you think will benefit from working together, or spend some time on work you feel is needed, perhaps sharing strategies for doing some of the mathematics in the games.

Organizing Folders The end of Session 4 marks the end of Investigation 2. This would be a good time for students to bring their mathematics folders up to date, checking that they have completed and saved Student Sheets 1 through 6.

Sessions 3 and 4 Follow-Up

Homework

Playing Hidden Coins Send home the game directions, How to Play Hidden Coins. Students enjoy challenging family members—especially adults— to try their skill at picking out coins without looking. Ask them to report back to class on how successful these new players were, and if there are some coins that seem to confuse people more than others.

Teacher Note

About Choice Time

Choice Time is an opportunity for students to work on a variety of activities that focus on similar content. The activities are not sequential; as students move among them, they continually revisit important concepts and ideas, such as grouping in tens and hundreds to estimate or add. Some activities require that students work in pairs, while others can be done alone or with a partner. Many involve some type of recording or writing; these records will help you assess students' growth.

Students can use the Choice List (p. 140) to keep track of their work. As students finish a choice, they write it on their list and attach any written work they have done. Some teachers list the choices for each day on the board or overhead and have students copy the list at the beginning of class. Students are then responsible for checking off completed activities. You may also want to make the choices available at other times during the day.

In any classroom, there will be a range of how much work students can complete. Each choice may also provide extensions and additional problems for students to do once they have completed their required work. Choice Time encourages students to return to choices they have done before, doing another page of problems or playing a game again. Students benefit from such repeated experiences.

If you and your students have not used a structure like Choice Time before, establish clear guidelines when you introduce it. Discuss what students' responsibilities are during Choice Time:

- Try every choice at some time.
- Be productively engaged during Choice Time.
- Work with a partner or alone.
- Keep track, on paper, of the choices you have worked on.
- Keep all your work in your math folder.
- Ask questions of other students when you don't understand or feel stuck.

Some teachers establish the rule, "Ask two other students before me," requiring students to check with two peers before coming to the teacher for help. You may need to try organizing Choice Times in a couple of different ways and decide from experience which approach best matches the needs of your students.

Collaborating with the Authors

Teacher Note

Every unit in this curriculum is a guide, not a prescription or recipe. We tested these activities in many different classrooms, representing a range of students and teachers, and revised our ideas constantly as we learned from students and teachers. Each time we tried a curriculum unit in a classroom, no matter how many times it had been tried and revised before, we discovered new ideas we wanted to add and changes we wanted to make. This process could be endless, but at some point we had to decide that the curriculum worked well enough with a wide range of students.

We cannot anticipate the needs and strengths of your particular students this particular year. We believe that the only way for good curriculum to be used well is for teachers to participate in continually modifying it. Your role is to observe and listen carefully to your students, to try to understand how they are thinking, and to make decisions, based on your observations, about what they need next. Modifications to the curriculum that you will need to consider throughout the year include:

- changing the numbers in a problem to make the problem more accessible or more challenging for particular students
- repeating activities with which students need more experience
- engaging students in extensions and further questions
- rearranging pairs or small groups so that students learn from a variety of their peers

Your students can help you set the right pace and level of challenge. We have found that, when given choices of activities and problems, students often do choose the right level of difficulty for themselves. You can encourage students to do this by urging them to find problems that are "not too easy, not too hard, but just right." Help students understand that doing mathematics does not mean knowing the answer right away. Tell students often, "A good problem for you is a problem that makes you think hard and work hard—and you might have to try more than one way of doing it before you figure it out."

The *Investigations* curriculum provides more than enough material for any student. Suggestions are included for extending activities, and some curriculum units contain optional sessions (called Excursions) to provide more opportunities to explore the big mathematical ideas of that unit. Many teachers also have favorite activities that they integrate into this curriculum. We encourage you to be an active partner with us in creating the way this curriculum can work best for your students.

Using Number Patterns

What Happens

Sessions 1 and 2: The 300 Chart Students enter numbers on a partially filled 300 chart, looking for patterns. They use their completed charts to compute differences. They then play a bingo game that involves adding and subtracting multiples of ten on a 101 to 200 chart.

Session 3: Related Problem Sets Students work on closely linked groups of problems, called Related Problem Sets, involving addition, subtraction, and money. They meet in groups to assess their own work, and students individually write down something they learned from the group.

Sessions 4 and 5: Addition and Subtraction Strategies During two sessions of Choice Time, students choose from Related Problem Sets, 101 to 200 Bingo, and other games learned earlier in this unit. As an assessment task, they do number and money problems based on the games.

Mathematical Emphasis

- Using known answers to find others
- Subtracting on a 300 chart and with a calculator
- Adding and subtracting multiples of ten

What to Plan Ahead of Time

Materials

- Calculators: at least 1 per pair (Sessions 1–2)
- Scissors, tape (Sessions 1–2)
- Numeral Cards from previous investigations: 1 deck per pair (Sessions 1–2, 4–5)
- Game markers (such as cubes, square tiles, counting chips): 2 per student (Sessions 1–2)
- Colored pencils, crayons, or markers (Sessions 1–2, 4–5)
- Buckets or boxes of interlocking cubes (Session 3)
- *Alexander, Who Used to Be Rich Last Sunday,* by Judith Viorst (Macmillan, 1989) (Sessions 4–5, optional)
- Overhead projector (Sessions 1–2)

Other Preparation

- Duplicate student sheets and teaching resources (located at the end of this unit) in the following quantities. If you have Student Activity Booklets, copy only the transparencies marked with an asterisk.

  For Sessions 1– 2

  Student Sheet 8, 300 Chart (p. 115): 1 per student

  How to Play 101 to 200 Bingo (p. 126): 1 per student, homework

  101 to 200 Bingo Board (p. 127): 3–4 per student

  Tens Cards (p. 128): 1 set per pair; 1 transparency* of each (cut apart); 1 set per student, homework

  Numeral Cards* (p. 136): 1 transparency of each (cut apart)

  For Session 3

  Student Sheet 9, Related Problem Sets (p. 118): 1 packet per student. Staple together the pages to make a six-page Related Problems Sets booklet for each student. For second-language learners, use coin stamps or refer them to their Coin Value Strips to identify the coins on pages 4–6 of these booklets.

  For Sessions 4–5

  Student Sheet 10, Numbers and Money (p. 124): 1 per student

  101 to 200 Bingo Board (p. 127): 1–2 per student

  Choice List (p. 140): 1 per student (optional)

Sessions 1 and 2

The 300 Chart

Materials

- Student Sheet 8 (1 per student)
- Scissors, tape
- Game markers (2 per student)
- Calculators (at least 1 per pair)
- 101 to 200 Bingo Board (3–4 per student)
- Numeral Cards (class sets from earlier sessions)
- Tens Cards (1 set per pair, and 1 set per student, homework)
- Colored pencils, crayons, or markers
- Transparencies of Numeral Cards and Tens Cards (cut apart)
- Overhead projector
- How to Play 101 to 200 Bingo (1 per student, homework)

What Happens

Students enter numbers on a partially filled 300 chart, looking for patterns. They use their completed charts to compute differences. They then play a bingo game that involves adding and subtracting multiples of 10 on a 101 to 200 chart. Their work focuses on:

- writing numbers from 1 to 300
- counting by 10's and multiples of 10
- finding differences between numbers
- using a calculator to subtract

Ten-Minute Math: Exploring Data To introduce students to recording and analyzing data, do the Ten-Minute Math activity Exploring Data three or four times before the end of this unit. Remember that Ten-Minute Math activities are designed to be done outside of regular math time.

To start the activity, you or the students choose a question about themselves. Because this is a 10-minute activity, the data they collect must be something they already know or can observe easily in the class. For example:

How did you get to school?

How many hours of TV did you watch yesterday?

How are your shoes fastened: laces, Velcro, buckle, no fastener?

Are you wearing red in any of your clothing? purple? yellow?

❖ **Tip for the Linguistically Diverse Classroom** Selecting data questions that can be demonstrated by modeling or can be easily drawn with rebuses will help ensure that all students understand the activity.

As students supply their individual pieces of data, quickly graph the data using a line plot, list, table, or bar graph. Ask students to describe what they see in the data, generate new questions, and, if appropriate, make predictions about how the data might be different if they were to ask the same question another time.

Students will be doing more extended investigations of data in other grade 4 units *(The Shape of the Data,* and *Three out of Four Like Spaghetti)*, but will benefit from the quick, repeated practice that this Ten-Minute Math activity offers.

For a full description of this activity, see pp. 93–94.

Activity

Filling in the 300 Chart

Distribute the three pages of Student Sheet 8 to each student (unstapled). Students trim off the bottom of each sheet and tape them together to make a long, continuous 300 chart. (Help them find a way to fold the three parts together so the chart can be put away neatly.)

Once students have their charts put together, write on the board two or three numbers from each page of the chart; for example, 34, 129, 67, 222, 185, 291, 281, 12, 102. Students write in these numbers where they belong. Then each student in turn suggests another number for everyone to write in. After everyone has called for one number, students continue filling their charts, using any patterns they see to complete rows and columns.

Observing the Students While students are working, watch how they handle the 300 chart:

- Are students using patterns to fill in the numbers?
- Do they recognize counting by 10's on the columns?
- Are they able to move across 100 and 200 without difficulty?

Activity

How Many Steps?

Introduce this activity to pairs as they finish filling in their charts. Pairs will need their completed 300 chart, two game markers each, and calculators to find differences between numbers.

To start, each student names a number. (As you are introducing the task, pick two numbers reasonably near each other, such as 132 and 146.) Both students place markers on both numbers on their 300 charts and figure out the distance from one number to the other.

Watch for students who include both numbers in the count ("132, 133, 134, 135, 136—that's five") instead of counting steps ("132 to 133, that's 1, 134, 135, 136—that's four").

When they have both decided on the difference between the numbers, one student uses the calculator to compute the difference by subtracting one number from the other. If students subtract the larger number from the smaller, the calculator will show a negative answer. If this occurs, suggest that they subtract the smaller number from the larger and compare that result with the first.

How are the two answers different? How are they the same? Why is that?

Observe to see how students compute differences. Encourage them to count by 10's or to figure out multiples of 10. Look for efficient ways students compute using the 300 chart; for example, to find the distance from 134 to 182, they might "add 50 to get 184, subtract 2, that's 48."

Once or twice, stop the class to do a problem together and share ways of figuring it out. As students explain their approaches, write down the steps they follow. You might use this idea to start a classroom list of ways to find differences on the 300 chart.

Activity

Counting by 10's

In preparation for the game 101 to 200 Bingo, give students practice, through unison counting, adding on 10's to numbers that are not multiples of ten.

We're going to do some counting together by 10's. We'll begin with a way that I *know* you can count. I want you to count together in a whisper. Start with 10, and count by 10's: 10, 20 ...

Count with students, but don't get ahead of them. All will likely do fine until after 100, when some may say 200. Stop there, and ask students to take a moment to think and to talk quietly with each other about what number follows 100 when counting by 10's, and what number comes after that. Write 80, 90, 100 on the board. When students agree, write 110, then continue to count on in unison to over 200.

Then pose a harder problem:

This time, we'll start with 3 and count on by 10's. Take a moment to decide what the next number is, and the number after that. [*Pause.*] OK, let's start with 3 and add on 10's. Whisper the numbers: 3, 13 ...

This time, write the numbers on the board as students say them. When you have your back turned to the class, stop saying the numbers yourself, and listen to the students. When you hear several different numbers for the same step, write them all on the board, and ask students to think again about which one comes next. Continue your count beyond 100.

If students are doing well, pose more difficult problems like these:

Start at 84, and count on by 10's.
Start at 394, and count on by 10's.
Start at 83, and count backward by 10's.
Start at 127, and count backward by 10's.
Start at 20 (or 23 or 30), and count on by 20's.

Activity

Playing 101 to 200 Bingo

Before you begin this activity, students will need to prepare the sets of Tens cards, which provide positive and negative multiples of ten from 10 to 70. Hand out the sheets of Tens Cards and scissors. Each pair cuts apart one set for their use in the game, making a deck of 40 cards (including 6 Wild Cards). If you think students will not have scissors at home, they could also cut apart sets for home use at this time.

For this game, each pair will need a single copy of the 101 to 200 Bingo Board; a set of Numeral Cards (from earlier sessions); a deck of Tens Cards; and colored pencils, crayons, or markers.

101 TO 200 BINGO BOARD

101	102	103	104	105	106	107	108	109	110
111	112	113	114	115	116	117	118	119	120
121	122	123	124	125	126	127	128	129	130
131	132	133	134	135	136	137	138	139	140
141	142	143	144	145	146	147	148	149	150
151	152	153	154	155	156	157	158	159	160
161	162	163	164	165	166	167	168	169	170
171	172	173	174	175	176	177	178	179	180
181	182	183	184	185	186	187	188	189	190
191	192	193	194	195	196	197	198	199	200

© Dale Seymour Publications®   127   *Investigation 3 • Resource*
Mathematical Thinking at Grade 4

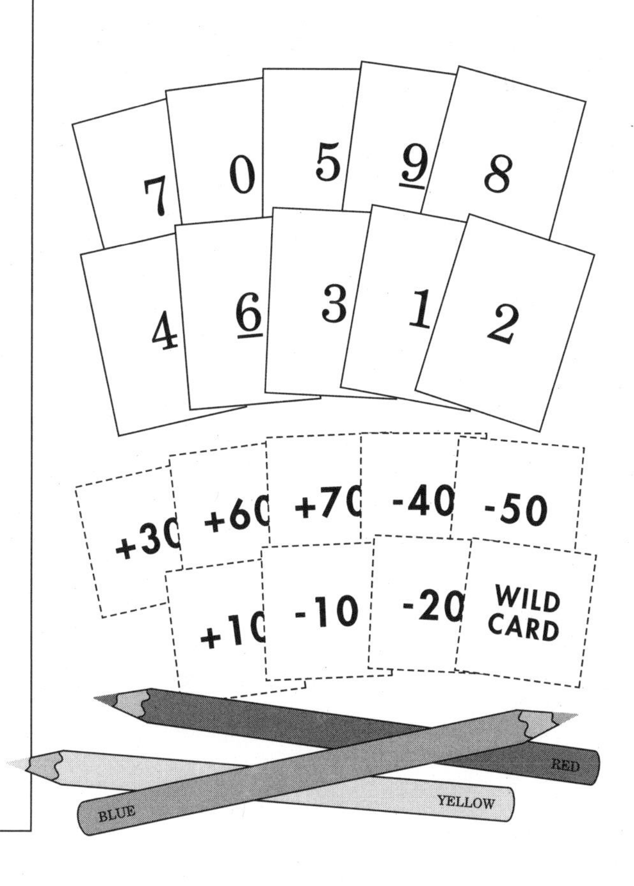

Introduce this game to the whole class by modeling it on the overhead projector with transparent Numeral Cards and Tens Cards, or introduce it to a few pairs at a time with students using their own Numeral Cards and Tens Cards. If you introduce it to small groups, enlist those who have learned the game to help you teach other students.

Before the game, each player starts by taking a 1 from the Numeral Card deck. (Two more 1's are left in the deck.) Players keep this card to use throughout the game. They then shuffle both the Numeral Cards and Tens Cards and place each pack face down on the table.

Players work together on a single 101 to 200 Bingo Board, taking turns and helping each other with their turns. To determine a play, one player draws two Numeral Cards (these will be used as digits) and one Tens Card. The play arranges the 1 and the two other numerals to make a number between 100 and 199, then adds (or subtracts) the number on the Tens Card, and circles the resulting number on the board.

The goal is for the players, working together, to circle five adjacent numbers in a row, column, or diagonal. As necessary, demonstrate the meaning of *adjacent*—with sides or corners touching.

Note: Some combinations will yield only numbers that are not on the 101 to 200 Bingo Board. When this happens, students should make up their own rules about what to do. For example, one class let the player take another turn. Another class decided that the Tens cards could be *either* added or subtracted in that instance. Another class allowed the player to use the 1 card anywhere in the number—a rule that is useful in some specific cases.

Students will have chances to play this game for homework and in later sessions in this investigation.

Sessions 1 and 2 Follow-Up

Homework

Playing 101 to 200 Bingo Students teach the game 101 to 200 Bingo to family members or friends. Each student will need How to Play 101 to 200 Bingo, copies of the 101 to 200 Bingo Board, and a set of Tens Cards. They should still have the Numeral Cards that they took home during Investigation 2.

Session 3

Related Problem Sets

Materials

- Related Problem Sets booklet (1 per student)
- Interlocking cubes

What Happens

Students work on closely linked groups of problems, called Related Problem Sets, involving addition, subtraction, and money. They meet in groups to assess their own work, and students individually write down something they learned from the group. Their work focuses on:

- comparing addition or subtraction problems to see similarities and differences
- doing addition and subtraction mentally
- consulting with other students

Activity

Solving Sets of Related Problems

Before you hand out the Related Problem Set booklets, introduce the idea of Related Problem Sets at the board.

For the next few days, we'll be thinking about good strategies for figuring out harder addition and subtraction problems. We'll be looking at Related Problem Sets. Each problem set is a group of four, five, or six problems that are related to each other in some way. You can work on the problems in a problem set in any order. You can use cubes to help solve them, but not the calculator.

There's one important thing about these Related Problem Sets: You are to find the answers to them without writing anything down. Then, after you find the answers, you will write down how you thought about the problems.

Present the following two problem sets on the board, the first subtraction, the second addition:

13 – 5 =
23 – 5 =
43 – 5 =
103 – 5 =
203 – 5 =

4 + 8 =
24 + 8 =
54 + 8 =
94 + 8 =
194 + 8 =
254 + 8 =

Students complete the two problem sets, working on the problems within a set in any order. They may copy the problems as necessary to keep track of their answers, but otherwise should be working in their heads (or with cubes) rather than on paper. As they finish, small groups of two or three students get together to compare strategies and answers. As the students are working and comparing solutions, circulate and take note of the strategies they are using and any errors they make.

When everyone has answers to both problem sets, ask students to share with the whole class strategies they have figured out or have learned from each other. You might focus their thinking with questions like these:

Did you learn something useful from someone else? Tell us about it.

In what order did you solve the problems?

Did anyone solve them in a different way?

Did you learn anything from one problem that helped you solve another problem? How are the problems similar?

Activity

Teacher Checkpoint

Writing About Strategies

Following the discussion, hand out the prepared booklets of Related Problem Sets. Explain that students will be doing these problem sets over the next few days. Today, they are to start on the first few pages.

Looking at one problem set at a time, students solve any problems that they can mentally and write only the answers on the page, then use those answers to solve the other problems in that problem set. Students choose one problem set on each page to write about. In the space next to that problem set, they write any patterns they noticed or how they used answers to some problems to solve others. Students who are interested may write about both problem sets.

❖ **Tip for the Linguistically Diverse Classroom** Students who are not writing comfortably in English may show their thinking with numbers and with symbols (circles and arrows), showing the patterns they followed.

Observe to see how students are writing about their solutions. If they don't know what to write, ask them to tell you how they thought about a specific problem or two. If they can express their ideas orally, tell them to write down just what they told you.

Students are just learning to write about their thinking, so their writing will be less rich than their thinking and their oral descriptions. In observing and talking with them, notice:

- Are students able to work out problems mentally?
- Do students use what they learn from one problem to help them solve another? Are they using patterns to do the problems? Do they reflect on their own learning?
- Can students subtract across 100 (105 – 7, 205 – 7)?
- Can they see a similarity between 105 – 7 and 15 – 7?

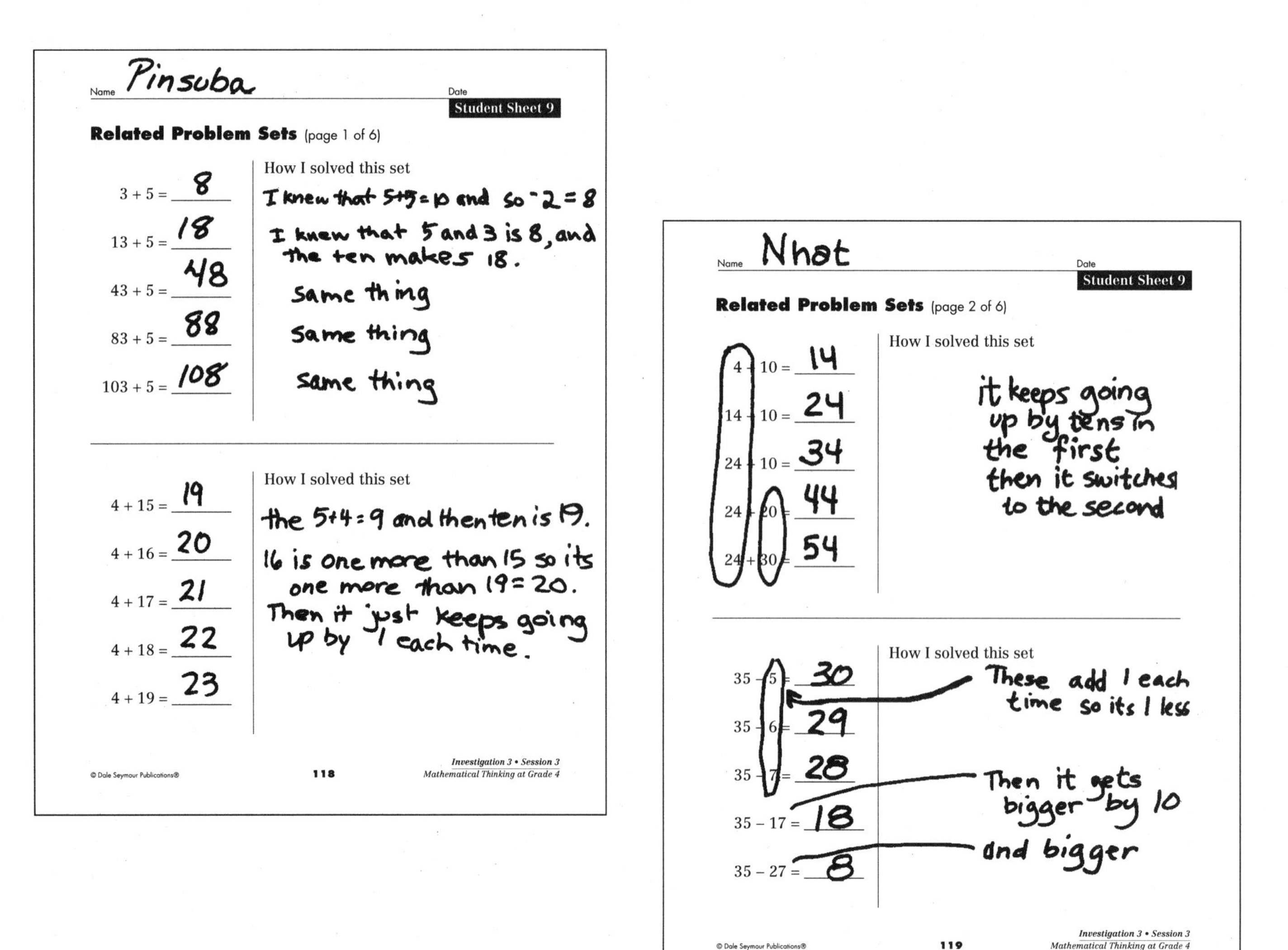

Name Pinsuba Date

Student Sheet 9

Related Problem Sets (page 1 of 6)

3 + 5 = 8
13 + 5 = 18
43 + 5 = 48
83 + 5 = 88
103 + 5 = 108

How I solved this set

I knew that 5+5=10 and so -2=8
I knew that 5 and 3 is 8, and the ten makes 18.
same thing
same thing
same thing

4 + 15 = 19
4 + 16 = 20
4 + 17 = 21
4 + 18 = 22
4 + 19 = 23

How I solved this set

the 5+4=9 and then ten is 19.
16 is one more than 15 so its one more than 19=20.
Then it just keeps going up by 1 each time.

© Dale Seymour Publications® 118 *Investigation 3 • Session 3 Mathematical Thinking at Grade 4*

Name Nhat Date

Student Sheet 9

Related Problem Sets (page 2 of 6)

4 + 10 = 14
14 + 10 = 24
24 + 10 = 34
24 + 20 = 44
24 + 30 = 54

How I solved this set

it keeps going up by tens in the first then it switches to the second

35 – 5 = 30
35 – 6 = 29
35 – 7 = 28
35 – 17 = 18
35 – 27 = 8

How I solved this set

These add 1 each time so its 1 less
Then it gets bigger by 10
and bigger

© Dale Seymour Publications® 119 *Investigation 3 • Session 3 Mathematical Thinking at Grade 4*

Activity

Comparing Solutions in Groups

After all students have completed a few pages of Related Problem Sets, group them in twos and threes to discuss their strategies.

This is a time for learning from one another. Talk among yourselves about how you worked out the answers for each set. Do not change your answers unless you are absolutely convinced the other answer is correct.

After you compare your ways of doing the problems, I will ask you what you learned from talking with the people in your group. Did you learn a different way to think about a problem? Did you find that you made the same kind of mistake more than once?

While students are talking, observe whether everyone is participating. In each small group, ask the students whether they have learned anything that will help them do the rest of the Related Problem Sets. Point out the space at the bottom of each page where they are to write about something they learned from doing the problems, or from correcting their mistakes, or from comparing their work to that of other students.

❖ **Tip for the Linguistically Diverse Classroom** Pair second-language learners with English-proficient students to check problems. Allow students to respond to the question at the bottom of each page orally. Encourage them to refer back to their problems by pointing or underlining key parts as they communicate their thoughts to you.

Working at their own speed, students continue to work on the Related Problem Set booklets and write about their strategies. Some may finish all the problems in this session; most will not complete them until Choice Time in Sessions 4 and 5. If students finish quickly, check that they have written clear explanations.

Session 3 Follow-Up

Homework

Related Problem Sets Depending on the pace of student work in class, you might ask students to do one page of Related Problem Sets as homework. Remind them to write about how they did each set. If they take home their Related Problem Sets booklets, they must remember to bring them back the next day for continued work on them during Choice Time.

Sessions 4 and 5

Addition and Subtraction Strategies

Materials

- Choice List (1 per student, optional)
- Students' Related Problem Sets booklets
- 101 to 200 Bingo Boards (1–2 per student)
- Numeral Cards (class decks)
- Tens Cards (decks from earlier sessions)
- Colored pencils, crayons, or markers
- *Alexander, Who Used to Be Rich Last Sunday* by Judith Viorst (optional)
- Close to 100 Score Sheets (optional)
- Coin Cards, plastic coins (optional)
- Student Sheet 10 (1 per student)

What Happens

During two sessions of Choice Time, students choose from Related Problem Sets, 101 to 200 Bingo, and other games learned earlier in this unit. As an assessment task, they do number and money problems based on the games. Their work focuses on:

- mental addition and subtraction
- adding and subtracting multiples of 10
- managing their time
- working with other students

Activity

Choice Time: Learning Together

Planning for Choice Time For the next two sessions, you may set up from two to five activity choices to be going on simultaneously in the classroom. Students work independently and in pairs and threes for all of Session 4. At the beginning of Session 5, you will present the assessment task on Student Sheet 10, Numbers and Money. As students finish work on the assessment, they return to the Choice Time activities for the rest of the session.

You may want to open Session 4 with just the first two or three of these choices, and add additional choices in Session 5 if students are finishing their Related Problem Set booklets and have each played a few rounds of 101 to 200 Bingo.

How to Set Up the Choices If you set up your choices at centers, show students what they will find at each center. Otherwise, make sure they know where they will find the materials they need.

Choice 1: Related Problem Sets—students' own booklets of Related Problem Sets, pencils

Choice 2: 101 to 200 Bingo—Numeral Cards, Tens Cards; 101 to 200 Bingo Boards; colored pencil, crayon, or marker for each player

Choice 3: How Much Alexander Spent (optional)—copy of the Judith Viorst book *Alexander, Who Used to Be Rich Last Sunday,* pencils, notebook paper

Choice 4: Close to 100 (optional)—Numeral Cards, copies of Close to 100 Score Sheet, pencils

Choice 5: Collecting Dollars (optional)—Coin Cards, plastic coins

Choice 3 is dependent on having a copy of the Judith Viorst book. This is the only new activity that will need introduction; see the description (following) for details. Review the other choices with the class as necessary.

Choice 1: Related Problem Sets

Students may continue to work on their Related Problem Set booklets. As before, they should complete a whole set at a time and write about how they used one or more problems to help solve the others. When they finish two sets, they compare their work with that of a classmate. Then, at the bottom of the page, they write something they learned from this comparison.

Ask students *not* to use calculators on these problems, because they should be using strategies based on their number sense.

Choice 2: 101 to 200 Bingo

Students may play 101 to 200 Bingo in pairs or groups of three. Remember that this is a cooperative game played on a single Bingo Board; players take turns and help each other with their turns. The object is to circle five adjacent numbers vertically, horizontally, or diagonally. Review the rules (p. 51) as necessary.

Choice 3: How Much Alexander Spent

If you have a copy of the book *Alexander, Who Used to Be Rich Last Sunday,* read it aloud to the class, perhaps outside of math time. Explain that the prices of things may seem surprisingly low because the story took place a long time ago, when everything cost a lot less.

As a Choice Time activity, students work in pairs or groups of three. They look through the book to add up all the money that Alexander spent. They may want to keep track on paper as they figure the total.

Choice 4: Close to 100

Review the rules to this game (p. 28) as necessary. Encourage students to try the positive and negative scoring variation, or to invent a new game or a variation of the game using the Numeral Cards or calculators.

Choice 5: Collecting Dollars

Review the rules for Collecting Dollars (p. 39) as necessary. Remind students of the variation—making combinations of Coin Cards that equal *any* whole number of dollars. Challenge students to invent other variations, or to create other Coin Cards with differing totals.

Observing the Students

While students are working together, observe and talk with them to see whether they are learning from one another.

- In 101 to 200 Bingo, are they learning new strategies from their partner, or do they just accept help to get the turn played well?
- While doing Related Problem Sets, are they learning by checking with another student, or are they just erasing and putting in the other student's answers?

You might take some whole-group time toward the end of Session 4 to ask students to tell about some of the things they have learned from working with a partner. List these with the names of both students. You could begin this list with instances you learned of in your conversations with students.

Activity

Assessment

Numbers and Money

At the beginning of Session 5, hand out Student Sheet 10, Numbers and Money, to assess the work students have done up to this point in the unit.

❖ **Tip for the Linguistically Diverse Classroom** You may need to do this assessment orally with individual students who have limited English proficiency. Read aloud each problem, modeling actions whenever necessary to ensure comprehension.

Observe students while they are working. See the **Teacher Note,** Assessment: Numbers and Money (p. 62), for suggestions on what to be looking for, both while students are working and when you are evaluating their papers later.

Some students may need most of Session 5 to complete this assessment task. Those who finish early may quietly continue to do Choice Time activities.

Sessions 4 and 5 Follow-Up

Complete the Booklet Students who have not yet completed their Related Problem Sets booklets finish them as homework. If they have already finished the Related Problem Sets booklet, they continue to play one of this unit's math games at home.

Invent a New Game You could challenge students to invent a new game that involves numbers or money. They may want to incorporate some of the cards or gameboards used for other games in this unit, or they might create new ones. Encourage them to bring their games in to share with the class. Plan to review any new games with the creator first to help work out any difficulties in the game rules.

Teacher Note **Assessment: Numbers and Money**

The work of Irena, Vanessa, and Kyle represents the range of answers you might see on the assessment tasks on Student Sheet 10, Numbers and Money.

1. Here are 6 Numeral Cards:

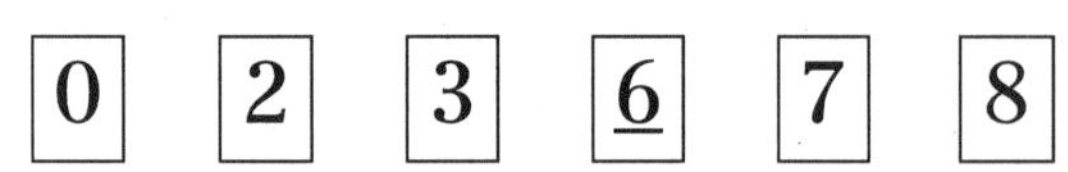

Use four of the numbers. Make two 2-digit numbers that you can add to make a number close to 100. Can you get 100 exactly? How close can you get?

Irena made 100 exactly by combining the cards into 68 + 32. Other students will make 100 with 62 + 38. Most students will get either 100 or very close to 100 (78 + 23 = 101; 23 + 67 = 90; 62 + 37 = 99).

Vanessa added incorrectly (86 + 20 = 100); she may have been concentrating on making 100 with the tens digits and forgot the ones digits. Other students sometimes use two of a certain digit when only one is available (78 + 22 = 100; 63 + 37 = 100).

Irena wrote the problem horizontally and added in her head, showing strong mental computation skills.

Both Vanessa and Kyle, on the other hand, wrote their two-digit numbers one under the other and added traditionally, with carrying.

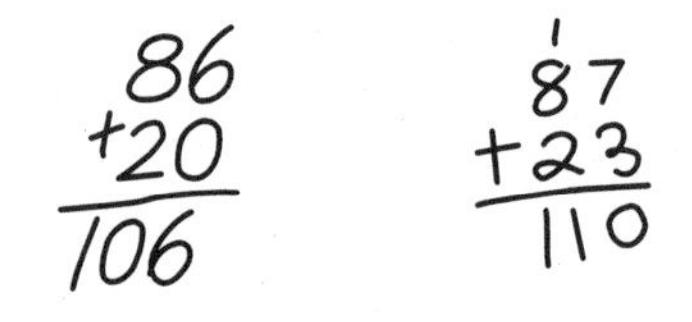

As you observe students working, when you find them writing the numbers in columns, ask them to show you how they can add in their heads.

A few students may not estimate in order to plan a combination. They will pick numbers randomly and test them by doing the traditional algorithm. When you see this happening, encourage estimation by asking them first to make *any* number with two of the digits, and then tell you about how large a number they need to make 100.

2. Write the answers to these counting problems in the blanks:

a. Start with 58. Count up by 10's by adding 10 each time.

58 __ __ __ __ __ __ (+ 10 each time)

b. Start with 4. Count up by 20's.

4 __ __ __ __ __ __ (+ 20 each time)

c. Start with 137. Count *backward* by 10's.

137 __ __ __ __ __ __ (– 10 each time)

d. What advice would you give to someone who was trying to count by 10's and 20's?

Most students will be successful at adding tens. Some, however, may have difficulty with adding by twenties and subtracting tens.

Irena's three counts were all perfect, and her advice for counting by 10's and 20's showed a corresponding grasp of the task: "Add 1 or 2 to the tens. Or take away 1."

Vanessa counted by 20's this way:

4 24 44 64 84 104 114

She has the overall pattern, but runs into difficulty working above 100.

Kyle recognized correctly that the changes for counting by 20's are only in the tens digits, but after the first answer, he reverted to the easier counting by 10's.

4 24 34 44 54 64 74

Kyle also counted up by 10's, instead of backwards, for the last problem.

A few students will make several mistakes in a series, apparently from counting by ones without looking for a pattern in either the tens or the ones digits: 4, 24, 33, 54, 64, 75, 84. These students need lots of practice adding or subtracting multiples of 10. They might benefit from playing a simpler version of bingo, using the 100 chart instead of the 101 to 200 Bingo Board. Limit the Tens Cards at first to +10, –10, +20, and –20; they need to see these relationships in two-digit numbers.

3. Count and make groups of coins.

a. What is the total value of these coins?

b. What is the total value of 3 quarters, 2 dimes, 2 nickels, and 2 pennies?

On problems 3a and 3b, Irena worked mentally and just wrote down a total. She shows the ability to count money and use coins in a practical way.

Vanessa wrote number amounts next to each coin picture (3a) and drew in coins next to the number values (3b) before adding. Kyle listed the values in columns to add:

$$\begin{array}{r} 75 \\ +20 \\ \hline 95 \end{array} \qquad \begin{array}{r} 95 \\ +10 \\ \hline 1.05 \end{array} \qquad \begin{array}{r} 1.05 \\ .02 \\ \hline 1.07 \end{array}$$

Even though students like Vanessa and Kyle may find the correct total, they are not yet working with coins in a useful way.

A few students may still not recognize the coins shown. Let them compare the pictures with real coins, if possible.

c. Show what coins you could use to make 63¢.

d. Show a different way to make 63¢.

Irena's answer starts with larger denominations and adds smaller ones as needed, in the most efficient use of different coins:

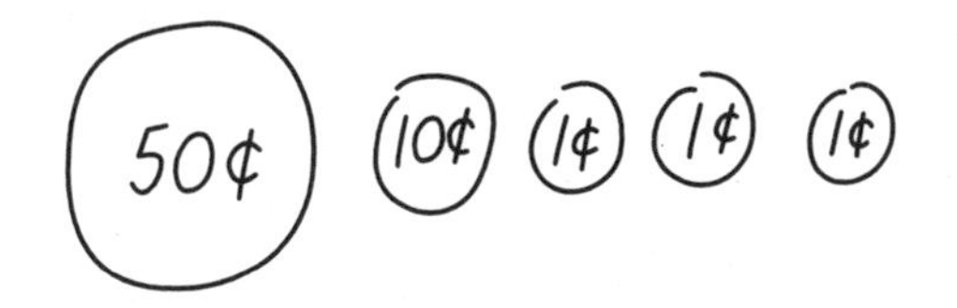

She uses her answer in 3c for 3d, exchanging two quarters for the half dollar and two nickels for the dime.

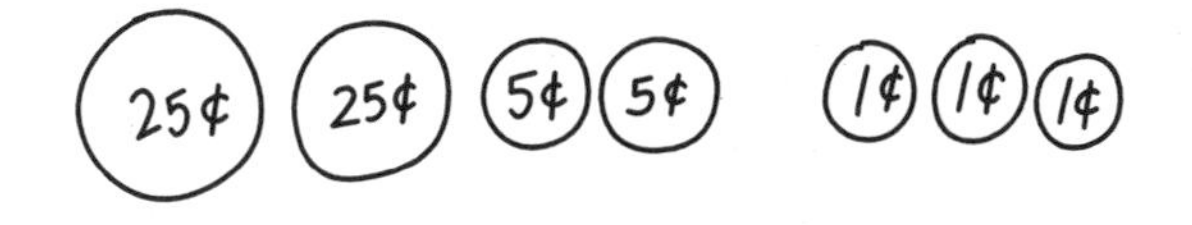

Vanessa started out on the right foot, with a half dollar, but then depended on 13 pennies to take her to 63¢.

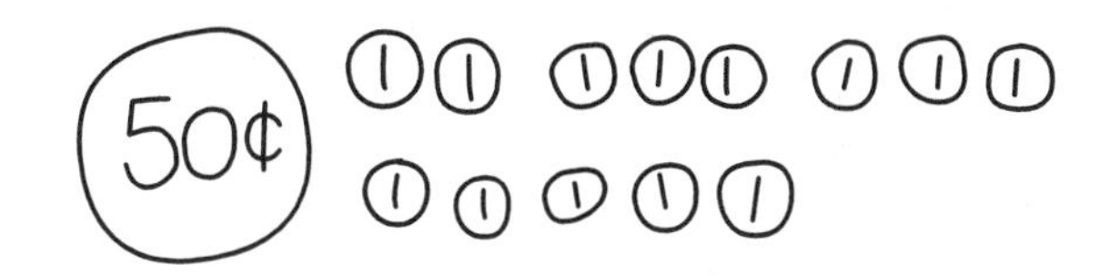

For 3d, she also replaced the half dollar with two quarters, and substituted a nickel for five of the pennies, but still used 8 pennies to reach the total.

Kyle seemed lost in problems 3c and 3d, and worked out his answers with plastic coins and the teacher's help.

When students are not yet comfortable with finding the total values of coins, suggest that they practice this frequently, maybe at home, or for a real purpose in school—counting milk money, for example. Show them efficient techniques, such as starting with the coins that have the largest value, and grouping the coins in piles that equal 25¢, 50¢, or a dollar.

Making Geometric Patterns

What Happens

Session 1: Patterns with Mirror Symmetry Using pattern blocks, pairs of students make designs that have a line of symmetry, and distinguish them from designs with no clear pattern. They also look for examples of symmetry in the classroom or outdoors environment.

Session 2: Patterns with Rotational Symmetry Pairs of students construct pattern block designs that grow from a central hexagon and have rotational symmetry. The class discusses what makes a pattern a pattern.

Sessions 3 and 4: Patterns and Nonpatterns On a bulletin board display of patterns and nonpatterns, students post clippings brought from home along with the pattern block designs they have made. They play a game in which they try to build a design from oral descriptions. They write about one of the designs they made earlier, and work in groups to guess which description fits which design.

Sessions 5 and 6: Symmetrical Geoboard Patterns Using geoboards, students make designs with rotational symmetry. They also make a geoboard design with mirror symmetry, and copy half of their design onto dot paper for another student to complete. Students investigate the number of lines of symmetry in their designs. They finish the unit with an assessment, drawing one design with only mirror symmetry and one with only rotational symmetry, and writing about their work.

Mathematical Emphasis

- Distinguishing between geometric patterns and random designs
- Distinguishing between mirror symmetry and rotational symmetry
- Writing about designs

What to Plan Ahead of Time

Materials

- Pattern blocks: 1 bucket of 250 per 4–6 students (Sessions 1–6)
- Ruler or straightedge: 1 per pair (Session 1)
- Pencils, crayons, or markers in red, green, yellow, and blue (Sessions 1, 2, 5–6)
- Scraps of patterned fabric, wrapping paper, or wallpaper (Session 1, optional)
- Scissors: 1 per student (Session 2)
- Stick-on notes (Sessions 3–6)
- Geoboards with rubber bands in assorted colors: 1 per student or per pair (Sessions 5–6)
- Overhead projector
- Overhead pattern blocks (optional)

Other Preparation

- Prepare the pattern block sets by removing the narrow parallelograms and the squares, leaving hexagons, triangles, trapezoids, and wide parallelograms (diamonds), or plan to have students remove them in Session 1. Read the **Teacher Note,** Pattern Block Shapes (p. 70).
- Prepare a bulletin board where you can post student work and pictures or clippings they bring in to demonstrate patterns and nonpatterns. (Sessions 3–4)
- Duplicate student sheets and teaching resources (located at the end of this unit) in the following quantities. If you have Student Activity Booklets, copy only the transparency marked with an asterisk.

For Session 1

Triangle paper (p. 134): 3–4 per student (homework)

For Session 2

Student Sheet 11, Mirror Symmetry and Rotational Symmetry (p. 130): 1 per student

Triangle paper (p. 134): 1 per student

For Sessions 5–6

Student Sheet 12, Multiple Lines of Symmetry (p. 131): 1 per student (homework)

Shaded Geoboard Design* (p. 132): 1 transparency

Geoboard Dot Paper (p. 133): 6 per student (4 for class, 2 for homework)

Triangle paper (p. 134): 4 per student (2 for class, 2 for homework)

Session 1

Patterns with Mirror Symmetry

Materials

- Pattern blocks (1 bucket per 4–6 students)
- Ruler or straightedge (1 per pair)
- Triangle paper (3–4 per student, homework)
- Pencils, crayons, or markers in red, green, yellow, and blue
- Overhead projector (optional)
- Overhead pattern blocks (optional)
- Scraps of fabric, wrapping paper, or wallpaper (optional)

What Happens

Using pattern blocks, pairs of students make designs that have a line of symmetry, and distinguish them from designs with no clear pattern. They also look for examples of symmetry in the classroom or outdoors environment. Their work focuses on:

- making patterns with mirror symmetry
- working cooperatively with a partner
- copying a pattern onto paper

Ten-Minute Math: Exploring Data Continue to do Exploring Data as a Ten-Minute Math activity outside of math time. During this investigation, you might collect data related to the numbers of each pattern block that students use as they create designs. Small groups could be assigned to total the number of a given shape—for example, triangles—in all the designs.

Quickly graph the data using a line plot, list, table, or bar graph. Ask students to describe what they see in the data and make predictions about whether the data might be different if they were to gather it another time.

Alternatively, students in pairs can take turns choosing a question and collecting and recording the data. Or you could post a table or graph form on which students can record their own data any time during the day, and then discuss the gathered data as a whole class at the end of the day.

For a full description and variations on this activity, see pp. 93–94.

Activity

Symmetry with Pattern Blocks

Distribute the buckets of pattern blocks to groups. If you haven't removed the narrow parallelograms and squares, students can do that now. Provide a plastic bag or other container to store these extra blocks.

For this activity, creating pattern block designs with a line of symmetry, each student will need a sheet of triangle paper. Have extras for students who have time to create more than one design. The first step is to darken one of the horizontal or diagonal lines near the middle of the page to use as a line of symmetry. This is done easily with a straightedge.

If you have an overhead projector, use it to demonstrate as you explain the activity. If not, use pattern blocks on a table or the floor where students can gather around to see. Make a line of symmetry by drawing a line or placing a stick. Students will enjoy participating in the demonstration on the overhead. If you don't have overhead pattern blocks, use regular pattern blocks and slightly separate adjacent blocks so students can see their outlines. A small mirror will be useful for supporting your demonstration of mirror symmetry.

Explain the procedure for this activity as follows:

You will be working in pairs to make symmetrical designs with pattern blocks. You'll be making more than one design, but you'll do each one together, working on one sheet of triangle paper until the design is finished.

Start with a sheet of triangle paper, marked with a line of symmetry. One side will be person 1's side, the other will be person 2's side. As you make your design, you will only work on your own side of the line.

- **Person 1 starts by putting a pattern block on the triangle paper, touching the line of symmetry along one side.**
- **Person 2 now puts the same kind of block in the mirror-image position on the other side of the line.**
- **Person 2 now gets to place a new block. Each block put down must touch either the line of symmetry or another block, at least by a corner if not a whole side.**
- **Person 1 puts a block in the mirror-image position.**

Continue to take turns like this. Use 12 blocks in all. You should each have three turns to place a new block in the design.

After each pair finishes one design, they leave those pattern blocks in place on the paper. Then they use 12 new pattern blocks to make a different design on the other sheet of triangle paper.

When they have finished both designs, partners can each color one design by lifting the blocks one at a time. Tell them to use colors that match the colors of the blocks. You may want to wait until pairs have finished both designs before you distribute the colored pencils, crayons, or markers.

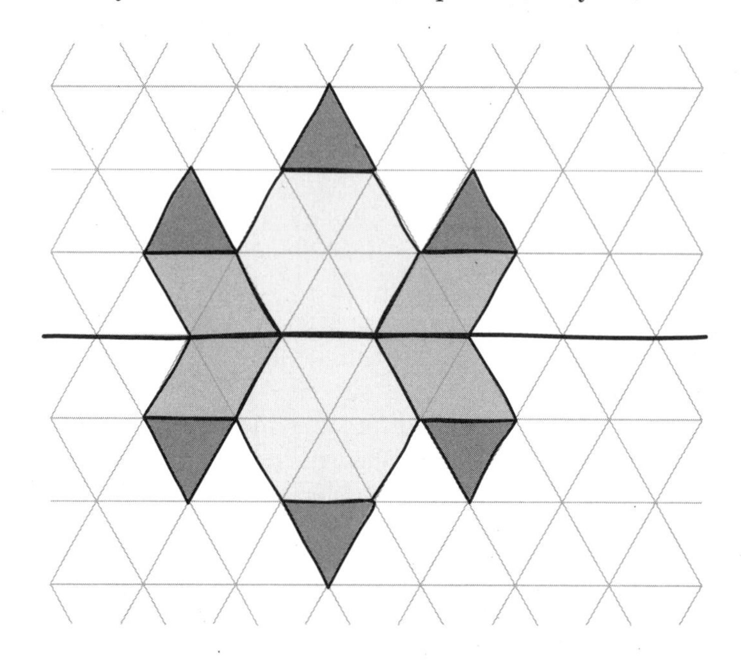

A Design Without Symmetry If there is time, students can use pattern blocks and triangle paper to make designs that have no symmetry. They could make a picture or an abstract design, but it must *not* be the same on one side as the other.

Save the designs from this activity—both with and without symmetry—for posting on the bulletin board in Sessions 3–4.

Activity

Symmetry in the Environment

Allow a few minutes at the end of the session for students to look in the classroom or outside to find examples of symmetry. If you have scraps of fabric, wrapping paper, or wallpaper, these are often good sources. On the board, list the things they notice.

Ask students to indicate with their hands the lines of symmetry in each shape. Some designs may have more than one line of symmetry; some may have rotational or mirror symmetry.

If someone finds a design with rotational symmetry, begin a discussion of the difference between mirror and rotational symmetry to prepare students for the activities in Session 2. See the **Teacher Note,** Pattern and Symmetry (p. 71), for information about these types of symmetry.

Session 1 Follow-Up

Homework

Finding Examples of Symmetry To collect items for your Patterns/No Patterns bulletin board in Sessions 3 and 4, ask students to bring in examples of symmetry they find, perhaps in advertisements, photographs, pieces of gift wrap, wallpaper, and fabric. Students may also make a list of other items they see outside of school that have symmetry but that can't be brought to class. Watch for some designs, such as clocks, where it is not clear whether or not there is symmetry. These will be especially useful for discussion.

Making a New Design Send home triangle paper for students to make a new design, with or without symmetry. Save their work for posting on the Patterns/No Patterns bulletin board.

Extension

Making Symmetrical Linear and 3-D Designs Some students may enjoy making linear designs (as can be seen in some beaded necklaces), starting in the middle, placing shapes on either side, and maintaining symmetry around the middle.

If you have or can borrow building blocks for the classroom, students enjoy building three-dimensional structures that have symmetry and balance—some have made very elaborate structures with perfect symmetry.

Teacher Note

Pattern Block Shapes

The pattern block set is made up of six shapes: a hexagon, a square, a triangle, two parallelograms (rhombuses), and a trapezoid. In most sets, each shape comes in one color: the hexagons are yellow; the triangles are green; the squares are orange; the trapezoids are red; the narrower parallelograms or rhombuses (which your students will probably call diamonds) are tan; and the wider parallelograms or rhombuses are blue (although in some sets, they are so dark as to appear purple).

Pattern blocks are related mathematically in terms of side length and angle measures. For example, six of the triangles can be put together to make one hexagon; two of the trapezoids make one hexagon. Your students will see some of these relationships readily. Because of these relationships, pattern blocks are useful for exploring many aspects of mathematics, including fractions, symmetry, and angles of polygons. They fit together to make beautiful and mathematically interesting patterns.

You may want to explore the pattern block relationships further by trying the following problem yourself:

> If the hexagon is one unit of area, what are the relative sizes of the other pieces?

While some of the relationships of the pieces are straightforward, you might find others challenging. Students who have done the *Investigations* grade 3 Fractions unit, *Fair Shares,* will be familiar with the use of pattern blocks to represent fractions, especially thirds and sixths.

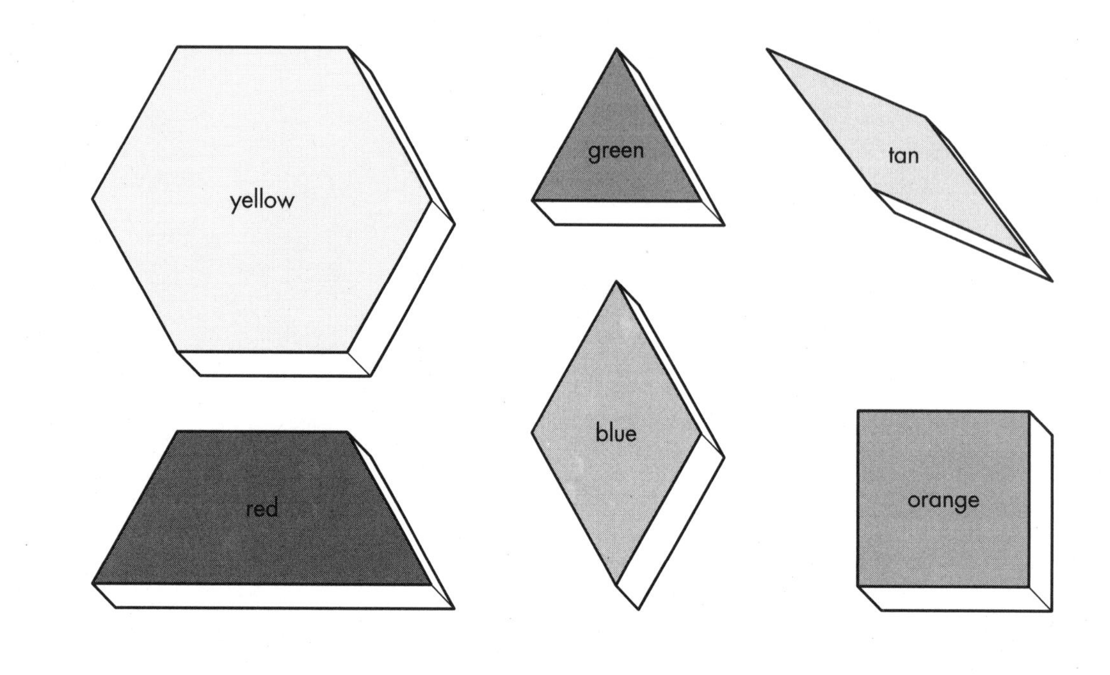

Pattern and Symmetry

Teacher Note

A pattern creates a regularity that can be copied. When we speak of patterns in fabric, we may mean a design made from colors or textures. Plaids have a color pattern; corduroys have a texture pattern. A pattern in a mathematical sense has a relationship among its component parts—sometimes in number, sometimes in relationship among data, sometimes in visual shape.

Number A pattern exists in each of these series of number relationships, making it possible to predict the next steps:

$10 \times 23 = 230$	$5 + 8 = 13$
$100 \times 23 = 2300$	$5 + 18 = 23$
$1{,}000 \times 23 = 23{,}000$	$5 + 28 = 33$
$10{,}000 \times 23 = ?$	$5 + 38 = ?$
$? = ?$	$? = ?$

1 3 6 10 ?

Data A fourth grade class that collected bedtime and rising time data in their school concluded that younger children sleep for more hours than older children. Although this pattern had some remarkable exceptions—such as one first grader who reported to sleep only five and a half hours a night—the information allowed them to make generalizations about the number of hours of sleep required by students at different ages, and to make predictions about the sleep patterns of children younger and older than those in their survey.

Shape The designs that follow have patterns that allow you to know what piece is missing. Patterns in which one half can be flipped or folded over onto the other half have mirror symmetry. The two halves are reflections of one another. The first three designs have mirror symmetry (if the question mark in each is replaced with the appropriate shape).

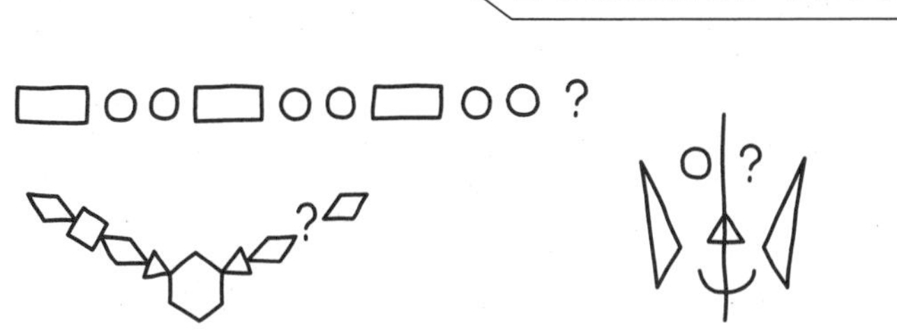

Mirror symmetry

Patterns that revolve around a central point have *rotational symmetry* or *circular symmetry*. A pattern with rotational symmetry can be turned a fraction of a circle—one-third, or 120° if it has three segments; one-fourth, or 90° if it has four segments; one-sixth, or 60° if it has six segments; one-twelfth, or 30° if it has twelve segments—and appear as though it hasn't moved at all. The two patterns below have rotational symmetry.

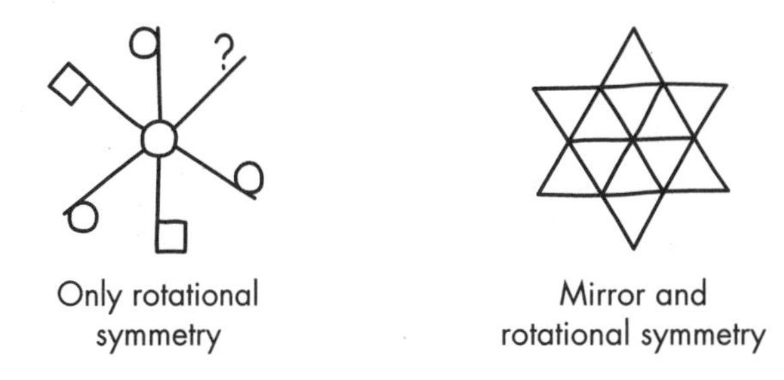

Only rotational symmetry

Mirror and rotational symmetry

A circle has infinite rotational symmetry; any turn will leave it looking the same. Some patterns with rotational symmetry—such as a circle, a regular hexagon like the yellow pattern block, any even-sided regular polygon, or the six-pointed star above—also have mirror symmetry. Others, such as the pinwheel design above and the trapezoid below, do not.

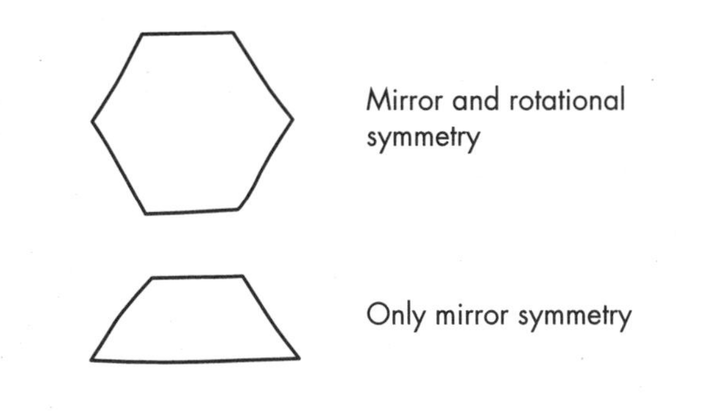

Session 2

Patterns with Rotational Symmetry

Materials

- Student Sheet 11 (1 per student)
- Scissors
- Buckets of pattern blocks
- Triangle paper (1 per student)
- Pencils, crayons, or markers in red, green, yellow, and blue

What Happens

Pairs of students construct pattern block designs that grow from a central hexagon and have rotational symmetry. The class discusses what makes a pattern a pattern. Their work focuses on:

- making patterns with rotational symmetry
- copying patterns onto paper
- thinking about other kinds of patterns

Activity

Two Symmetries: Which Is Which?

Hand out Student Sheet 11, Mirror Symmetry and Rotational Symmetry. Students may cut, fold, or manipulate the two designs in any way that helps them determine what kind of symmetry each demonstrates.

If students have trouble identifying which design has rotational symmetry, suggest that they hold each design on the table with a finger or a pencil in the center, and turn the design some part of a circle to see if there is a point where it appears the same as it did at the beginning. Emphasize that the placement of their finger must be in the center of the design.

How far must each design be rotated to see the same view more than once?

A design that must be rotated in a full circle before the same view can be seen again does *not* have rotational symmetry. A design that can be rotated less than a full circle to see the same view has rotational symmetry.

Try this with the design with rotational symmetry: How many times can you see the same view before you have turned it around a full circle?

A design with mirror symmetry can be divided in half so that each side is a reflection of the other. Remind students of the symmetrical designs they made in Session 1.

In the design with mirror symmetry, where is the line of symmetry?

Students may color the two designs in ways that show their symmetry.

As students continue to work with symmetry in this session, they may discover that they can make designs with *both* mirror and rotational symmetry in the same design.

Note: During this investigation, we use a number of specialized words in discussing symmetry. Don't insist that students use them, either in their speaking or, later, in their writing about their designs. What's important is that students find ways to communicate their ideas so others understand. See the **Teacher Note,** Using Mathematical Vocabulary (p. 76), for a discussion of how to use such terms in your class.

Activity

Patterns Around Hexagons

Distribute either triangle paper or plain paper to students, who will be working together to draw two designs with rotational symmetry. As when drawing their designs with mirror symmetry in Session 1, students work in pairs, cooperating to make the two designs. They work first with pattern blocks and then color the designs on their paper.

For each pattern, students begin with a hexagon shape, either using a whole yellow hexagon or constructing a *single-color* hexagon from trapezoids, triangles, or wide parallelograms. They place this in the center of their paper.

Using one type of block at a time, they build around the hexagon, arranging the blocks in a regular pattern. A limit of 15 or 16 blocks will help keep the designs simpler. Students may leave spaces to create a lacy effect, but the blocks must be placed equally around the hexagon to form a pattern with rotational symmetry.

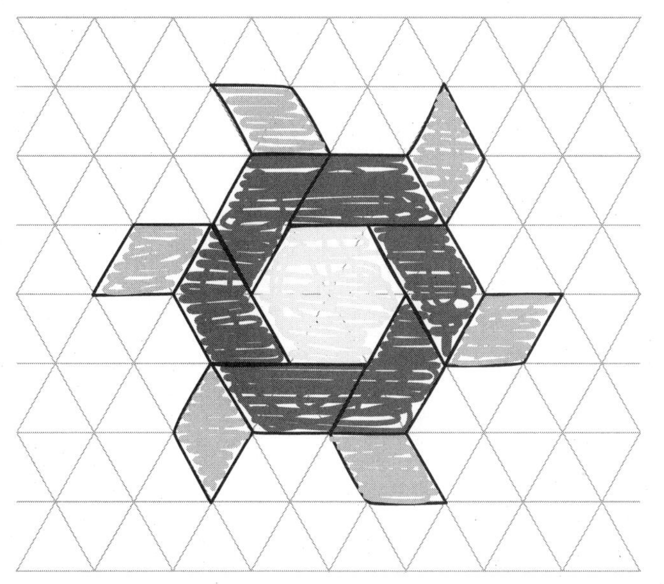

Checking for Rotational Symmetry As pairs finish their first design, students need to check to be sure that it has rotational symmetry.

Make a pencil mark at the bottom of one view of your design. Carefully rotate the triangle paper with the blocks on it, turning the paper in a circle around the center of the design. How far must you rotate it to see the same view of the design again?

If they need to rotate the paper a full turn (so that the pencil mark is at the bottom again), the design does not have rotational symmetry. Pattern block designs that do have rotational symmetry will need to be turned halfway, one-third of the way, or one-sixth of the way around to see the same view again.

Watch the pencil mark to keep track of how much you've turned the sheet. How much do you turn your design before it looks the same? Halfway around? One-third of the way? One-sixth of the way?

When each pair is sure their first design has rotational symmetry, they begin a second design. Again they start with a hexagon in the center, and use about 16 pattern blocks.

When students can confirm that they have two designs with rotational symmetry, they may record their work. Each student takes one design and, lifting one block at a time, colors the design on the paper, using colors that match the pattern blocks.

Activity

What Is a Pattern?

In the next session, students will be sorting collected items into patterns and nonpatterns for a classroom display. To prepare for this, allow 10 or 15 minutes at the end of this session to talk about pattern. Students should become familiar with the idea that a pattern allows us to predict the way something will repeat. We can cover up part of a visual pattern and know what is underneath.

You might show students examples of number or shape patterns and ask them to continue the patterns or to suggest others. See the **Teacher Note,** Pattern and Symmetry (p. 71), for some examples, or use patterns like the following:

Patterns in number series

3, 6, 9, 12, ...

1, 2, 4, 8, ...

Patterns that give clues to computation

$3 \times 10 = 30$

$7 \times 10 = 70$

$14 \times 10 = ?$

Patterns in the ones digits
(Related Problem Sets)

$3 + 5 = 8$

$23 + 5 = 28$

$14{,}583 + 5 = ?$

Patterns of growing shapes

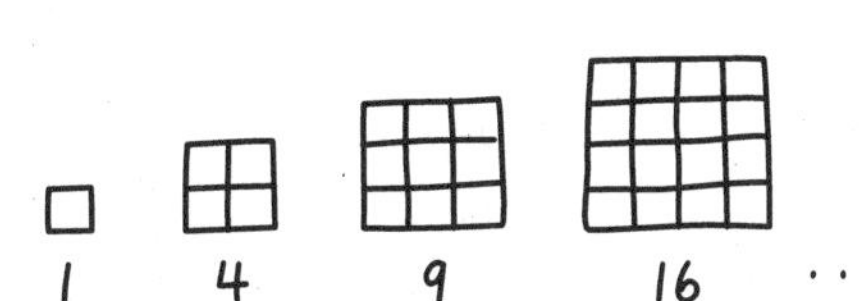

Patterns in a linear arrangement of shapes

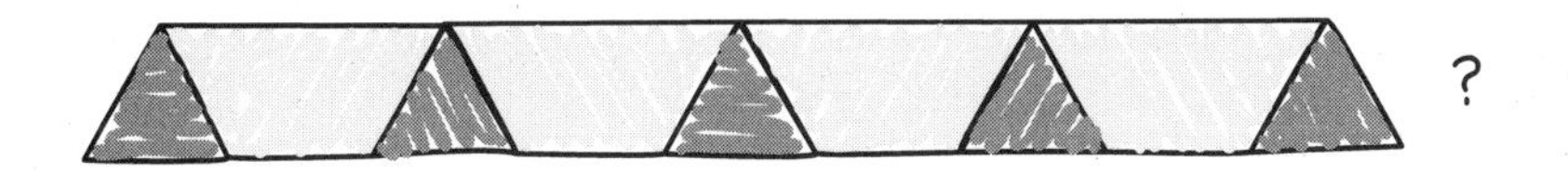

As you present examples, ask students to continue the patterns. Challenge them to design another pattern of the same type.

Session 2 Follow-Up

Collecting Designs for Display Students collect designs without symmetry for the classroom display. They may also finish designs they started in class that show mirror or rotational symmetry.

Pattern Puzzles Students might enjoy making up puzzles based on number or shape series. They draw or write their pattern, leaving space at the end for the patterns to be extended. Pattern puzzles could be posted on a *What's the Pattern?* bulletin board. Other students may extend the patterns by adding one number or shape in pencil; the originator may erase it if it isn't the correct next step. If students show interest in this, you might post the display in a hall to share it with other classes, challenging viewers to try extending the patterns.

Teacher Note — *Using Mathematical Vocabulary*

Students learn mathematical words the same way they learn other vocabulary—by hearing them used correctly, frequently, and in context. Young children learn words by hearing their families use language appropriately. When they make mistakes—"Look at the big doggie," says a young child pointing at a horse—adults use the correct words, and the child gradually learns the distinctions among all the four-footed animals.

We don't ask young children to memorize definitions of *horse* or *dog;* rather, we find opportunities for them to hear words being used in meaningful ways. Learning mathematical vocabulary is no different. Students do not learn how to speak mathematics by memorizing definitions, but by hearing these words frequently and having many opportunities to use them in context.

Use mathematical vocabulary accurately and frequently, and connect it with more familiar words that students may know:

Let's look at Rafael's pattern block design. Does it have *mirror symmetry?* Is one side a *reflection* of the other? How can you tell? Is there anything in the classroom that is *symmetrical?*

What can you say about the number of people wearing shoes with laces *compared to* the number of people wearing sandals? You think there are 12 more people wearing shoes with laces? How did you figure out that the *difference* is 12?

Throughout the *Investigations* curriculum, we will point out mathematical terms that are important for you to use in context. Don't insist that students use these terms. What is important is that they express their ideas and describe their strategies for solving mathematical problems clearly and accurately, using whatever words are comfortable for them.

As you use mathematical terms frequently in context, students will become used to hearing them and will begin to use them naturally. Even young children can learn to use mathematical vocabulary accurately when they hear it used correctly and in the context of meaningful activities.

DIALOGUE BOX

Discussing Symmetry

As the students in one class brought in from home their observations of things that have symmetry, the teacher took some time to discuss what they were finding.

Tell us some of the symmetrical things on your list that you saw but couldn't bring to class.

Kenyana: The smoke detector. It is round and you can cut it a few different ways and get it to be a mirror reflection.

B.J.: The curtains.

So there is one curtain on each side of the window, and they match each other?

B.J.: Yes.

Vanessa: A chair, but it matters how you split it. It is symmetrical only if you split the back and seat down the middle one way.

Marci: My cat.

Nick: No living thing is. I thought my dog was, but he's not because he has more spots on one side.

Marci: My cat is exactly the same on both sides.

Elena: The clock is symmetrical.

Is the clock symmetrical?

Dave: At 6 o'clock it is.

Marci: What about the second hand?

Vanessa: I don't think it is because of the numbers. If you took the numbers off, it might be symmetrical.

Marci: You'd have to take the hands off too. They are different lengths, and the second hand really gets in the way.

Does everybody agree that if we took the hands and numbers off, the clock would be symmetrical?

Nick: I have to think about it. It's hard for me to imagine it that way.

While you think about that, I have another question for you. If there is such a thing as a symmetrical cat, how many lines of symmetry does it have? How many ways could you cut it in half and have the same image on each side?

Tyrone: Only one way. You'd have to cut it between the ears and the long way down the tail.

That is an example of mirror symmetry. What else have we talked about that has mirror symmetry?

Marci: The chair, and probably the curtains.

Does anybody know what it means to rotate?

Nadim: To turn around?

I agree with that. Are there any items we mentioned that if you turn them a little bit you can get the same view as when you started? That is, are there any items that have rotational symmetry?

Kenyana: I think the smoke detector is like that. You can cut it a lot of different ways.

Marci: Maybe the clock without numbers and hands.

Tyrone: The chair?

If you were looking at a chair from above, how far would you have to rotate it to get the same view?

Tyrone: All the way around?

Nhat: A full circle.

An object has rotational symmetry when you can rotate it part way and it still looks the same. If you have to rotate it a full circle to get the same view, it doesn't have rotational symmetry.

Tyrone: But it might have mirror symmetry, right?

Sessions 3 and 4

Patterns and Nonpatterns

Materials

- Collected designs and clippings from home
- Stick-on notes
- Pattern blocks
- Overhead projector (optional)
- Overhead pattern blocks (optional)

What Happens

On a bulletin board display of patterns and nonpatterns, students post clippings brought from home along with the pattern block designs they have made. They play a game in which they try to build a design from oral descriptions. They write about one of the designs they made earlier, and work in groups to guess which description fits which design. Their work focuses on:

- recognizing patterns
- describing a visual design orally and in writing
- building a design from oral instructions

Activity

Teacher Checkpoint

A Display of Patterns

The class begins a bulletin board with designs and pictures displayed in two groups: those that have patterns and those that don't. Ask the students to suggest headings, such as *Patterns* and *Pictures Without Patterns.* Students can add to the display the patterns and nonpatterns they brought from home as well as any pattern block designs they have completed during this investigation.

Allow time for students to examine the display. Supply stick-on notes that they can put on any items they think are placed in the wrong group. As a class, talk about designs that are controversial.

Use this activity as a checkpoint. While students are posting their pictures, examining the display, and talking about items that are controversial, notice the following:

- Can they distinguish between designs that have patterns and those that don't?
- Can they make designs with symmetry?
- Can they see symmetry in other people's designs?
- Can they indicate the line of symmetry in a pattern with mirror symmetry?

- Can they point to the center of a design with rotational symmetry?
- Can they point out how a pattern has rotational symmetry but not mirror symmetry?

Many of these ideas are sophisticated. However, some students are more able to see these attributes of visual patterns than their teachers—including students who have difficulty doing accurate computation. At the same time, students who do well with number work may have difficulty in visualization tasks.

Activity

Writing About Our Designs

Students now select one of the pattern block designs that they have colored during this investigation and, on notebook paper, write a description of it. The description should tell which blocks (shapes) they used and how many of each, how their pattern grows, the number of lines of symmetry, and, if appropriate, what part of a circle they need to rotate the design to see the view they started with.

Pairs can work together on designs they created cooperatively, or they can each pick a different design to describe.

Since writing in mathematics may be new to your students, you may want to start them off with a sentence such as "Here is how we built our pattern block design." The description need only be three or four sentences, but insist that students include all the important information. Ask them to check their work this way:

If someone read your description, could that person pick out your design from all the others on display?

❖ **Tip for the Linguistically Diverse Classroom** Offer students the option of writing their description in their native language. If they do so, have them include rebus drawings over key words, for example, for the shapes they used. If they have not yet acquired written language skills, pair them with English-proficient students. They can provide drawings for each sentence.

Collect these designs and written descriptions for use in the activity at the end of Session 4, Guessing from Descriptions (p. 82).

Activity

Draw the Hidden Design

For this activity, students need to have buckets of pattern blocks within their reach. They will work in small teams and take turns being the leader. The leader makes an arrangement of pattern blocks, hidden from the view of others, and then describes orally how to make it. The rest of the group tries to replicate the hidden design.

First demonstrate the activity to the whole class. Build a simple design with three or four blocks, hidden from the view of the class (for example, put it on the overhead projector with the projector turned off, or on a desk with books propped around it). The design doesn't have to have a pattern, but it may be. (If you are working with real pattern blocks on the overhead, leave space between the pattern blocks so students will be able to clearly see each silhouetted shape when you show your design.)

Give the students an oral description of your design. Stop after each statement to give them time to build with their blocks what they think it looks like. Here is an example:

- **I put down a green triangle, with one point facing toward the left.**
- **To the right of the triangle, I put a blue diamond. One of its points is touching the middle of the side of the triangle.**
- **Then I put a red trapezoid above the other two blocks. One of its points touches the point where the green and blue blocks meet. It just fits.**
- **I put another red trapezoid block below the other blocks. The red blocks are mirror images of each other.**
- **There is a line of symmetry across the middle of my design. It looks like a flying creature with a green head, a blue body, and red wings.**

When you are finished with your directions, reveal your design. Hold a discussion about how clear the directions were. Ask students what you might have said to describe your design more clearly.

The object of this activity is for the leader and the rest of the team to collaborate—the leader is *not* trying to stump the others. The leader always tries to give a description that is as clear as possible; the rest of the team members listen carefully and ask good questions.

After you have demonstrated the activity, students divide into teams of three, or teams of four playing as pairs. The leader (or leader pair) for each team makes a hidden design. They might work behind a barrier, on a book or binder on one of their laps, or on a chair below the level of the desk or table where they're seated. The first design they try should have only 3 or 4 pieces.

When the design is finished, the leader describes it while the others try to follow the description and copy the design with pattern blocks. When they are finished, the design is revealed and compared with the copy.

Students take turns being the leader as they repeat the activity several times. If they are successful with three or four pieces, they can try a design with five or six pattern blocks.

❖ **Tip for the Linguistically Diverse Classroom** For this activity, group together students who speak the same native language. They can use their primary language to give the oral descriptions.

As you circulate, notice when the copied designs don't match the original. Encourage students to review the description that was given and try to understand what went wrong.

Activity

Guessing from Descriptions

In a new display, post only the designs students wrote about earlier. As students take turns reading their descriptions aloud, other students try to figure out which design is being described.

Doing this activity in small groups of five or six will help limit the number of choices. One teacher with a small class had each student come up to the front of the room and sit in "The Mathematician's Chair" while reading his or her description for others to guess the design.

Alternatively, you might post all the designs, each labeled with a number. Erase or cover any names that are written on the front of the designs. Compile and copy the written descriptions, *with* student names on them, and distribute these to the students. Students read each description and write the number of the matching design. Allow students to move around the room and consult with the authors of the descriptions for clarification.

❖ **Tip for the Linguistically Diverse Classroom** This activity should be comprehensible to everyone if the written descriptions included rebus drawings, as suggested on p. 79.

If it is too cumbersome for you to compile and copy all the descriptions, students could pass their written work around the class. Pairs of students could then work with two descriptions at a time, attempting to match them with the numbered designs.

My pattern looks like a diamond that would go on a ring. It is made up of mostly rhombuses. In the middle three triangles are connected. Connected to the triangles are red trapazoids. You have to turn it a third to get the same pattern. My pattern has no yellow hexagons. There are three triangles.

Sessions 5 and 6

Symmetrical Geoboard Patterns

What Happens

Using geoboards, students make designs with rotational symmetry. They also make a geoboard design with mirror symmetry, and copy half of their design onto dot paper for another student to complete. Students investigate the number of lines of symmetry in their designs. They finish the unit with an assessment, drawing one design with only mirror symmetry and one with only rotational symmetry, and writing about their work. Their work focuses on:

- making symmetrical patterns
- copying patterns from the geoboard to dot paper
- finding symmetry in other students' patterns

Materials

- Geoboards and rubber bands (1 set per student or per pair)
- Transparency of Shaded Geoboard Design
- Overhead projector
- Geoboard dot paper (6 per student; 4 for class, 2 for homework)
- Colored pencils, crayons, or markers
- Triangle paper (4 per student; 2 for class, 2 for homework)
- Stick-on notes
- Pattern blocks
- Student Sheet 12 (1 per student, homework)

Activity

Teacher Checkpoint

Symmetry on the Geoboard

Before you pass out rubber bands for the geoboard activities, discuss with students how to handle the rubber bands appropriately (for example, they are not to use them for shooting).

After distributing the geoboards and rubber bands, allow students some time to make designs and become accustomed to working with the geoboard. When students seem ready, tell them to put away all but four rubber bands for each geoboard. Students may work on this activity in pairs if you don't have enough geoboards for individual work.

This Teacher Checkpoint activity has two parts, one for rotational symmetry and one for mirror symmetry.

Rotational Symmetry

To introduce rotational symmetry on the geoboard, display the Shaded Geoboard Design on the overhead projector and ask students what kind of symmetry they see in it. (It has rotational symmetry and not mirror symmetry.) Rotate the design one-fourth (90°) for students to see that it looks the same from that viewpoint.

To demonstrate that the design does not have mirror symmetry, ask students to imagine a vertical line of symmetry through it.

SHADED GEOBOARD DESIGN

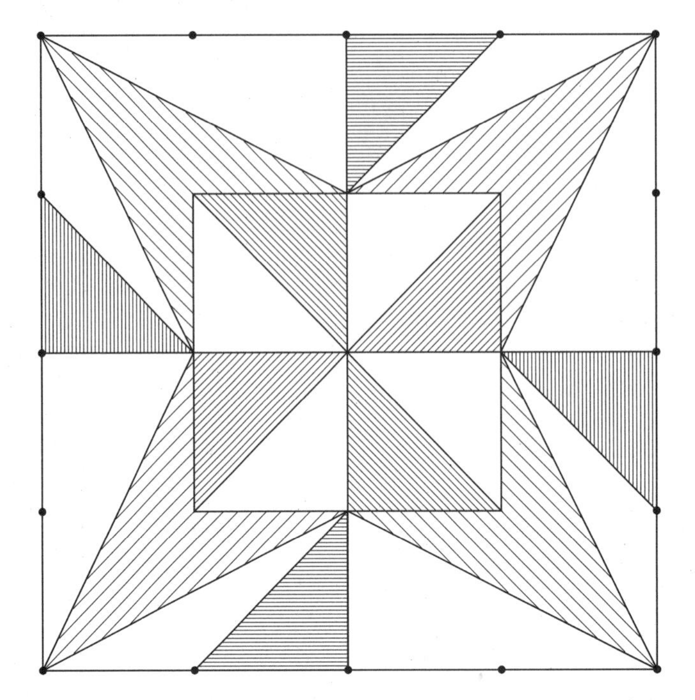

If you fold it over on this line, or imagine the reflection, what happens? Does the design on both sides of the line match up? Try a different line of symmetry—horizontal or diagonal. Does that work?

See if students agree that this design has only rotational symmetry. Then direct them to their geoboards.

Make a design on your geoboard that has rotational symmetry. Ignore the color of the rubber bands. Just use whatever colors you have.

Many students may make designs that have both mirror and rotational symmetry. It is particularly challenging to create a design on the square geoboard that has rotational symmetry but *not* mirror symmetry.

When students have completed their designs, they exchange them and look for the rotational symmetry in each other's designs. A good way to check is to actually turn the geoboards around the center point. If the students can do this and find the same view of the design in less than one full turn, the design has rotational symmetry.

While students are working, circulate to do a quick check of their designs. Ask some students (not only those who have made mistakes) to show you how their designs rotate.

Distribute geoboard dot paper for students to copy their designs. Allow time for students to go quietly around the room to look at the designs their classmates have made.

Mirror Symmetry

Students now use the geoboard to make mirror symmetrical designs. Distribute more geoboard dot paper for this part of the activity.

When students have completed a design that has mirror symmetry, they check the symmetry with a partner (or, if working in pairs, with another pair). When they are satisfied that they have a symmetrical pattern, they copy *half* of their design onto geoboard dot paper.

To clarify the idea of drawing only half of the design, draw a symmetrical geoboard design on the board and erase half of it, leaving only the dots.

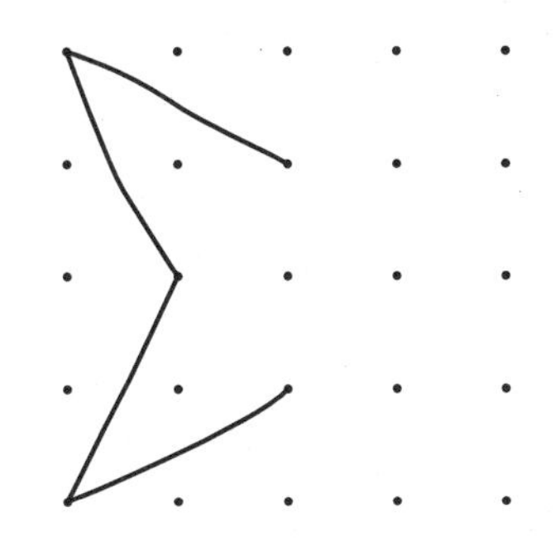

Students now trade half-designs with someone. (If two students have created a design together, they each make a copy of half of their design and trade with different students.) They then try to complete the design given to them. The geoboards are kept out of sight (or face down) during this step, and then brought out again when students are ready to check the accuracy of the completed drawing. The originator of the design corrects the drawing as necessary so that it has the symmetry that was planned.

Looking for Lines of Symmetry Students may color their designs for both rotational and mirror symmetry, then add them to the classroom display of designs with patterns. While students are coloring, ask some of them to show you the symmetry in their designs.

How many lines of symmetry do your designs have? Where are they?

Some students may have difficulty producing drawings that have mirror or rotational symmetry. However, all students should be able to do the following:

- Identify a design that has mirror symmetry and show that if you fold it, flip it, or look at the opposite sides, you can see that it is the same.
- Identify a design with rotational symmetry and explain how you can hold the center still and turn it a fraction of a whole circle to see that it looks the same from another view.

Check to be sure that students are aware of these differences in the two types of symmetry before they do the assessment at the end of the unit.

Activity

Counting Lines of Symmetry

Use the classroom display of finished designs as a focal point for more exploration. You might post the following questions near the display for students to consider:

- Which designs have only one line of symmetry?
- Which designs have four lines of symmetry?
- Do any of these designs have exactly two lines of symmetry?
- Which of the mirror designs also have rotational symmetry?

Students might use stick-on notes to indicate, near each design, how many lines of symmetry they find in it. Take some time for the class to discuss what they found out.

Do designs with four lines of symmetry always have rotational symmetry? How can you make a design that has mirror symmetry but not rotational symmetry?

Students can continue this investigation of lines of symmetry for homework, as suggested on p. 88.

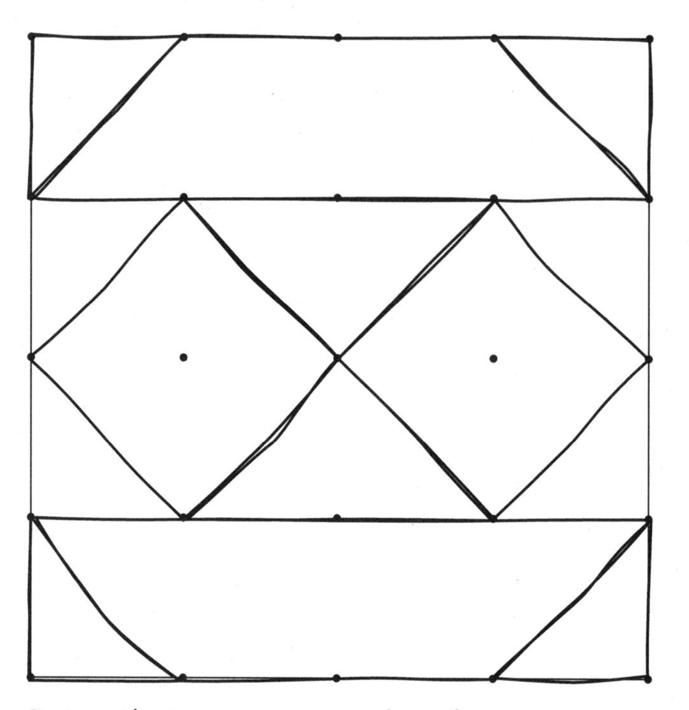

Design with mirror symmetry—two lines of symmetry. Also rotational symmetry.

Activity

Assessment

Mirror and Rotational Symmetry

There is no student sheet for this assessment, but students may want to use triangle paper or geoboard dot paper for their designs. Make these available. Students may also work on plain paper. They are also free to use pattern blocks or geoboards for planning their designs. Introduce the assessment task to the class.

Now I have a challenge for you. I want you to make two designs:

- **The first design must have mirror symmetry *but not* rotational symmetry.**
- **The second design must have rotational symmetry *but not* mirror symmetry.**

The hard part is making designs that have only one kind of symmetry, not the other. You may use pattern blocks or geoboards to work out your design and then copy it onto paper. Or, you may draw your design right on your paper.

When they have completed their drawings, the students write on the same page how they know that their design has rotational or mirror symmetry. Write these questions on the board for reference:

> How would you explain to someone else that your design has mirror symmetry?
>
> How would you explain to someone else that your design has rotational symmetry?

Collect student work for review. See the **Teacher Note,** Assessment: Mirror and Rotational Symmetry (p. 89), for guidelines on assessing student responses.

Activity

Choosing Student Work to Save

As the unit ends, you may want to use one of the following options for creating a record of students' work on this unit.

- Students look back through their folders or notebooks and write about what they learned in this unit, what they remember most, and what was hard or easy for them. You might have students do this work during their writing time.
- Students select one or two pieces of their work as their best work, and you also choose one or two pieces of their work, to be saved in a portfolio for the year. You might include students' written solutions to the assessments Numbers and Money (p. 61) and Mirror and Rotational Symmetry (p. 87). Students can create a separate page with brief comments describing each piece of work.
- You may want to send a selection of work home for families to see. Students write a cover letter, describing their work in this unit. This work should be returned if you are keeping year-long portfolios.

Sessions 5 and 6 Follow-Up

Homework

Multiple Lines of Symmetry After Session 5, send home Student Sheet 12 along with geoboard dot paper, triangle paper, or plain paper for students to make designs with different numbers of lines of symmetry. For homework, they:

Make one design with only one line of symmetry.

Make one design with two lines of symmetry.

Make one design with four lines of symmetry.

All of their designs will have mirror symmetry, but which have rotational symmetry and which do not? See if they can tell. They might add their designs to the class display for discussion at the beginning of Session 6.

Extension

Displaying More Patterns You might keep a bulletin board area all year to exhibit examples of patterns, both those made by students and those found ready-made. As students provide new patterns, take others down and return them. Include number patterns as well as visual designs.

Assessment: Mirror and Rotational Symmetry

Teacher Note

By the end of Investigation 4, most fourth grade students are able to make a design with mirror symmetry and a design with rotational symmetry. However, very few students that we observed made designs that had *only* rotational symmetry. Here is a beautifully simple example that one student produced:

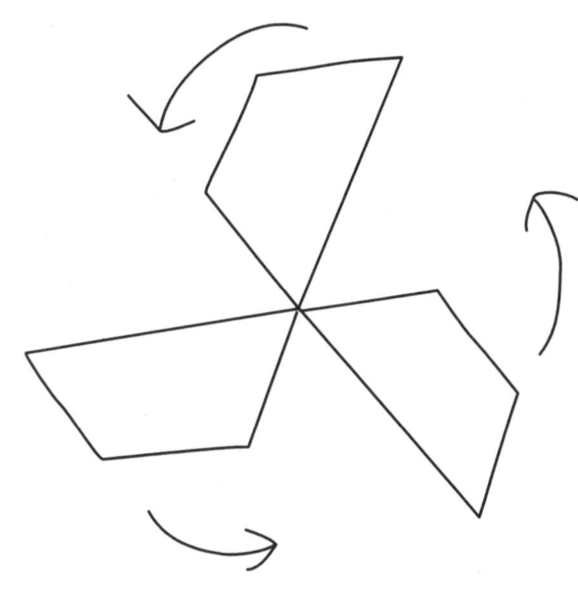

Many more students are able to make designs that have only mirror symmetry, like this one:

Some students will start with a line of symmetry (for a mirror design) or with a hexagon (for a circular design), then just add to the design until it has the kind of symmetry they are seeking. They then describe the symmetry they meant to draw, without considering whether the other kind of symmetry is also there. Thus, many ended up with designs that had both kinds of symmetry:

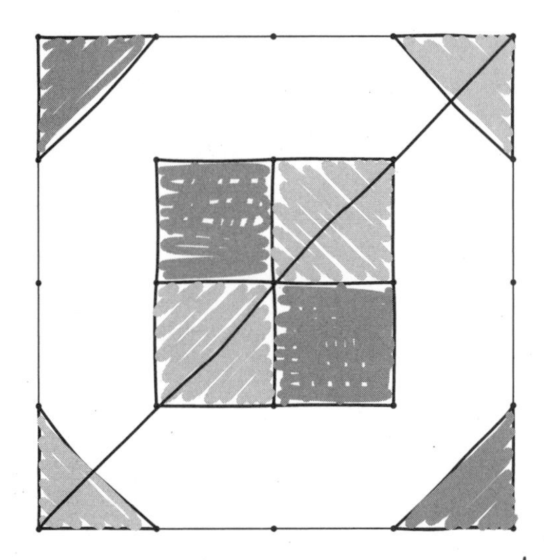

Mirror symmetry – It's the same on both sides of the line.

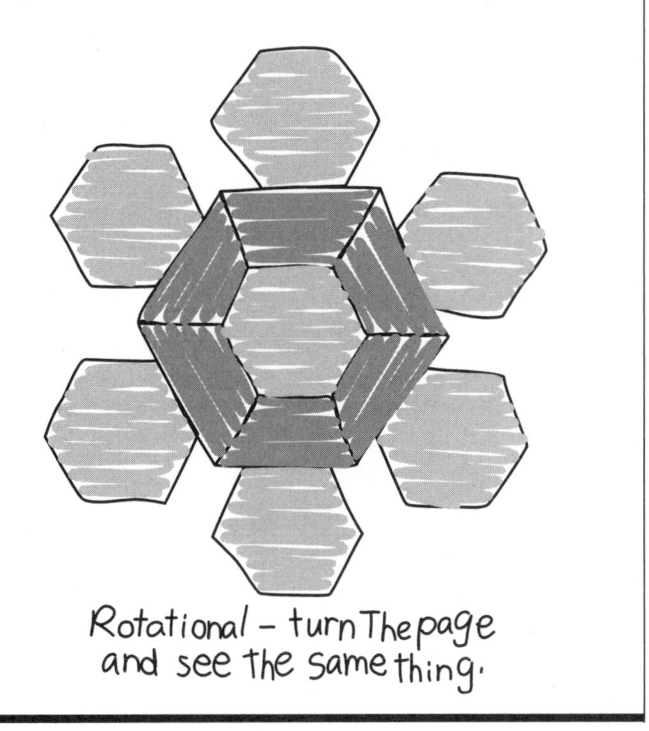

Rotational – turn The page and see the same thing.

Teacher Note

continued

A few students have difficulty with some parts of the task. They may show some idea of symmetry, but will make a design with one line of symmetry and say it has rotational symmetry. A student like this may be thinking of rotating the design around a point at its base, rather than a point in its center.

Other students will attempt a symmetric pattern but not execute it accurately.

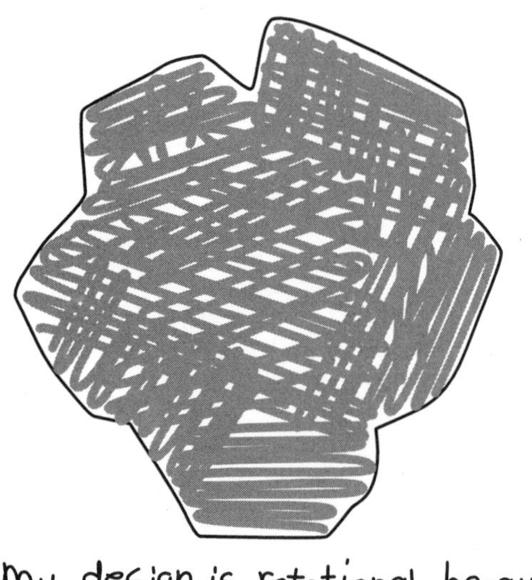

In assessing the written work, look to see how well students convey the basic characteristics of mirror and rotational symmetry. Here are some good reasons students have given for knowing why their designs have mirror symmetry:

> My design is symmetrical because it is a mirror.
>
> My design is symmetrical because if you fold it, it will end up on the same side.
>
> My design is symmetrical because I used the same shapes on both sides of my design.

And here are some clear reasons for knowing why a design has rotational symmetry:

> My design is rotational because you can turn it around and you can see the same thing.
>
> My design can turn around. It's spinning like a merry go round.
>
> My design is the same on every side when you rotate it.

If used as part of the suggested grade 4 *Investigations* sequence, this is the first unit of the year. This final investigation has given your students a chance to try some visual/spatial work, and for you to see which students are especially comfortable with this kind of work and which students have more difficulty. You may find, as we have, that throughout the year some students who have difficulty with number work will shine at spatial work. More visual/spatial work is offered in the units *Seeing Solids and Silhouettes* (3-D Geometry) and *Sunken Ships and Grid Patterns* (2-D Geometry). In addition, the unit *Different Shapes, Equal Pieces* takes a very visual approach to fractions.

Estimation and Number Sense

Basic Activity

Students mentally estimate the answer to an arithmetic problem that they see displayed for about a minute. They discuss their estimates. Then they find a more precise solution to the problem by using mental computation strategies.

Estimation and Number Sense provides opportunities for students to develop strategies for mental computation and for judging the reasonableness of the results of a computation done on paper or with a calculator. Students focus on:

- looking at a problem as a whole
- reordering or combining numbers within a problem for easier computation
- looking at the largest part of each number first (looking at hundreds before tens, thousands before hundreds, and so forth)

Materials

Calculators (for variation)

Procedure

Step 1. Present a problem on the chalkboard or overhead. For example:

9 + 62 + 91 + 30

Step 2. Allow 15 to 20 seconds for students to think about the problem. In this time, students come up with the best estimate they can for the solution. This solution might be—but does not have to be—an exact answer. Students do not write anything down or use the calculator during this time.

Step 3. Cover the problem and ask students to discuss what they know. Ask questions like these: "What did you notice about the numbers in this problem? Did you estimate an answer? How did you make your estimate?"

Encourage all kinds of statements and strategies. Some will be estimates; others may be quite precise:

"It's definitely bigger than 100 because I saw a 90 and a 60."

"It has to be 192 because the 91 and the 9 make 100 and the 30 and the 62 make 92."

Be sure that you continue to encourage a variety of observations, especially the "more than, less than" statements, even if some students have solved it exactly.

Step 4. Uncover the problem and continue the discussion. Ask further: "What do you notice now? What do you think about your estimates? Do you want to change them? What are some mental strategies you can use to solve the problem exactly?"

Variations

Problems That Can Be Reordered Give problems like the following examples, in which grouping the numbers in particular ways can help solve the problem easily:

6 + 2 – 4 + 1 – 5 + 4 + 5 – 2

36 + 22 + 4 + 8

112 – 30 + 60 – 2

654 – 12 + 300 + 112

Encourage students to look at the problem as a whole before they start to solve it. Rather than using each number and operation in sequence, they see what numbers are easy to put together to give answers to part of the problem. Then they combine their partial results to solve the whole problem.

Problems with Large Numbers Present problems that require students to "think from left to right" and to round numbers to "nice numbers" in order to come up with a good estimate. For example:

230 + 343 + 692

$3.15 × 9

8 + 1200 + 130

$5.13
$6.50
+ $3.30

Continued on next page

Present problems in both horizontal and vertical formats. If the vertical format triggers a rote procedure of starting from the right and "carrying," encourage students to look at the numbers as a whole, and to think about the largest parts of the numbers first. Thus, for the problem 230 + 343 + 692, they might think first, "About how much is 692?—700." Then, thinking in terms of the largest part of the numbers first (hundreds), they might reason: "300 and 700 is a thousand, and 200 more is 1200, and then there's some extra, so I think it's a little over 1200."

Fractions Pose problems using fractions and ask students to estimate the number of wholes the result is closest to. Start by posing problems such as $1/2 + 1/4$ or $1/2 + 3/4$, and ask, "Is the answer more than or less than one?" Eventually, you can include fractions with larger results and expand the question to "Is the answer closer to 0, 1, or 2?" Begin to include problems such as $5 \times 1/4$ and $3 \times 1/8$. Use fractions such as $9/4$, $50/7$, $100/26$, or $63/20$, and ask, "About how many wholes are in this fraction?"

Is It Bigger or Smaller? Use any of the kinds of problems suggested above, but pose a question about the result to help students focus their estimation: "Is this bigger than 20? Is it smaller than $10.00? If I have $20.00, do I have enough to buy these four things?"

Using the Calculator The calculator can be used to check results. Emphasize that it is easy to make mistakes on a calculator, and that many people who use calculators all the time often make mistakes. Sometimes you punch in the wrong key or the wrong operation. Sometimes you leave out a number by accident, or a key sticks on the calculator and doesn't register. However, people who are good at using the calculator always make a mental estimate so they can tell whether their result is reasonable.

Pose some problems like this one:

> I was adding 212, 357, and 436 on my calculator. The answer I got was 615. Was that a reasonable answer? Why do you think so?

Include problems in which the result is reasonable and problems in which it is *not*. When the answer is unreasonable, some students might be interested in figuring out what happened. For example, in the above case, you might say: I accidentally punched in 46 instead of 436.

Related Homework Options

Problems with Many Numbers Give one problem with many numbers that must be added and subtracted. Students show how they can reorder the numbers in the problem to make it easier to solve. They solve the problem using two different methods to double-check their solution. One way might be using the calculator. Here is an example of such a problem:

$$30 - 6 + 92 - 20 + 56 + 70 + 8$$

Exploring Data

Basic Activity

You or the students decide on something to observe about themselves. Because this is a Ten-Minute Math activity, the data they collect must be something they already know or can observe easily around them. Once the question is determined, quickly organize the data as students give individual answers to the question. The data can be organized as a line plot, a list, a table, or a bar graph. Then students describe what they can tell from the data, generate some new questions, and, if appropriate, make predictions about what will happen the next time they collect the same data.

Exploring Data is designed to give students many quick opportunities to collect, graph, describe, and interpret data about themselves and the world around them. Students focus on:

- describing important features of the data
- interpreting and posing questions about the data

Procedure

Step 1. Choose a question. Make sure the question involves data that students know or can observe: How many buttons are you wearing today? What month is your birthday? What is the best thing you ate yesterday? Are you wearing shoes or sneakers or sandals? How did you get to school today?

Step 2. Quickly collect and display the data. Use a list, a table, a line plot, or a bar graph. For example, a line plot for data about how many buttons students are wearing could look something like this:

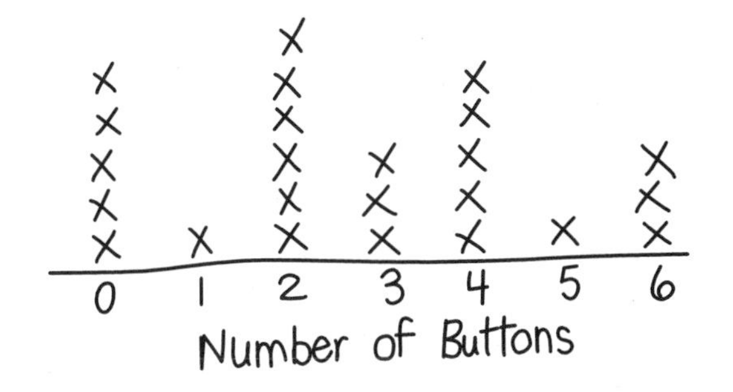

Step 3. Ask students to describe the data. What do they notice about it? For data that have a numerical order (How many buttons do you have today? How many people live in your house? How many months until your birthday?), ask questions like these:

"Are the data spread out or close together? What is the highest and lowest value? Where do most of the data seem to fall? What seems typical or usual for this class?"

For data in categories (What is your favorite book? How do you get to school? What month is your birthday?), ask questions like these: "Which categories have a lot of data? few data? none? Is there a way to categorize the data differently to get other information?"

Step 4. Ask students to interpret and predict. "Why do you think that the data came out this way? Does anything about the data surprise you? Do you think we'd get similar data if we collected it tomorrow? next week? in another class? with adults?"

Step 5. List any new questions. Keep a running list of questions you can use for further data collection and analysis. You may want to ask some of these questions again.

Variations

Data from Home For homework, have students collect data that involves asking questions or making observations at home: What time do your brothers and sisters go to bed? What do you usually eat for breakfast?

Data from Another Class or Other Teachers Depending on your school situation, you may be able to assign students to collect data from other classrooms or other teachers. Students are always interested in surveying others about questions that interest them, such as this one: When you were little, did you like school?

Continued on next page

Categories If students take surveys about "favorites"—flavor of ice cream, breakfast cereal, book, color—or other data that falls into categories, the graphs are often flat and uninteresting. There is not too much to say, for example, about a graph like this:

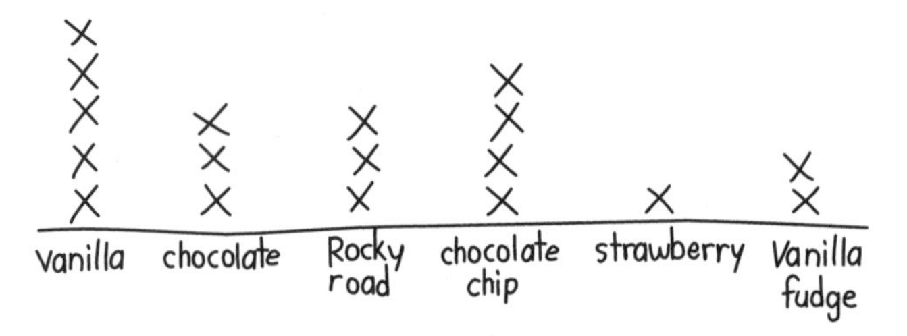

It is more interesting for students to group their results into more descriptive categories, so that they can see other things about the data. In this case, even though vanilla seems to be the favorite in the graph above, another way of grouping the data seems to show that flavors with some chocolate in them are really the favorites.

Chocolate flavors ~~||||~~ ~~||||~~ //

Flavors without chocolate ~~||||~~ /

The following activities will help ensure that this unit is comprehensible to students who are acquiring English as a second language. The suggested approach is based on *The Natural Approach: Language Acquisition in the Classroom* by Stephen D. Krashen and Tracy D. Terrell (Alemany Press, 1983). The intent is for second-language learners to acquire new vocabulary in an active, meaningful context.

Note that *acquiring* a word is different from *learning* a word. Depending on their level of proficiency, students may be able to comprehend a word upon hearing it during an investigation, without being able to say it. Other students may be able to use the word orally, but not read or write it. The goal is to help students naturally acquire targeted vocabulary at their present level of proficiency.

We suggest using these activities just before the related investigations. The activities can also be led by English-proficient students.

Investigation 1

the numbers 1 to 300

1. Create action commands that ask students to nonverbally identify numbers on the board.

   Put your finger on the number 96.

   Cover the number 225.

2. Use classroom items (paper clips, rubber bands) to help students count to 100.
3. Write the hundreds on the board: 100, 200, 300, 400, 500, 600, 700, 800, 900. Count by 100's with the students as you point to each number.
4. Challenge students to write different numbers that you call out.

   Write the number 279.

   Write the number 138.

   Write the number 205.

same, different, compare

1. Show students a set of Numeral Cards. Pick out two 3's. Trace the shape of each 3 with your finger, and tell students that when you compare these numbers, they look the *same*. Then pick out two 4's and say that these two are also the same. Ask the students to compare the cards and put into groups all those that are the *same*.
2. Pick out a 2 and a 9. Say that when you compare these numbers, you can see that they are not the same—they are *different*. As you select pairs of cards, ask students to compare them and tell you if they are the same or different.
3. Provide other items from the unit to compare—plastic coins, pattern blocks, interlocking cubes put together in simple shapes. Students should be able to identify items that are the same and items that are different.

add, plus, subtract, minus, difference

1. Give students different numbers of cubes, and ask them to count and report how many they have. Then move together two students' cubes into one pile.

   If I add Pinsuba's cubes and Tuong's cubes, how many will be in the new pile?

   Is the answer 15 because we added 7 + 8?

2. Write on the board the number problem that represents the adding of cubes you just did. Identify the plus sign.
3. Challenge students to add two different piles of cubes and find the total. Ask someone to write on the board the corresponding number problem.
4. Count out a pile of 12 cubes. Then remove 5 from the pile. Tell the students that when you subtract 5 cubes from the 12, you have 7 left. The *difference* between 12 and 5 is 7.
5. Write 12 – 5 = 7 on the board, and identify the minus sign.
6. Regroup the 12 cubes, and ask students to find the difference between 12 and 6. Have someone write and read the problem on the board.
7. Continue creating simple addition and subtraction problems with cubes, writing the number problems for each. When you are using the calculator in class, have students find the plus, minus, and equals keys.

lowest, close to

1. Write the following numbers on the board: 2, 9, 21, 39. Point to the 2 as you identify it as having the lowest value of these three numbers.
2. Ask students to find and mark 2, 9, 21, and 39 on the 100 chart. Point out that the 9 is *close* to

10. Point to 11, and say that 11 is another number that is close to 10.

Which of the numbers you marked is close to 20?

Which of the numbers is close to 1?

Which of the numbers is close to 40?

3. Write several groups of numbers on the board; for example, 102, 115, 126; 23, 35, 54; 287, 288, 289. Point to a number in one of the series and ask if it is the lowest number of that group. (It may or may not be.) If it is, students nod affirmatively; if it is not the lowest, students shake their heads negatively. Repeat with different numbers.
4. Indicating the numbers on the board, challenge students to find a number that is close to 100, a number that is close to 50, and a number that is close to 300.
5. Ask students to point to the lowest of all the numbers you wrote on the board.

Investigation 2

money: coins, cents, nickel, dime, quarter, dollar

1. Use a dollar, quarter, dime, and nickel—real, play money, or both—along with action commands to help familiarize students with these words.

   Put a quarter in your hand.

   Give me a dime.

   Put the nickel under the dollar.

   Put all the coins in a pile.

   Fold the dollar in half.

2. Create action commands that require students to identify coins by their value.

   Take a coin that is worth 5 cents.

   Give me a coin that is worth 10 cents.

   Show me which coin is worth the most money.

Investigation 3

pattern

1. Start a clapping pattern (hit knees, hit knees, clap; hit knees, hit knees, clap). Ask students to join in. Then change the pattern.

   Try this pattern with me.

   Who wants to show us a new pattern?

2. Give students two colors of interlocking cubes. Show them a pattern of cubes in a line, such as blue, red, red, blue, red, red. Ask students to make their own patterns.
3. Draw or write patterns on the board, identify them, and ask students to continue them.

   This is a pattern of shapes; what comes next?

   □ △ ○ □ △ ○ □ △ ○

   This is a pattern in numbers; what comes next? [*Write the numbers in order, then circle and say aloud each even number, pausing at 8.*]

   1 (2) 3 (4) 5 (6) 7 (8) 9 10 11 12 . . .

   This is another kind of number pattern; what comes next? [*Write the numbers in a column.*]

   3
   13
   23
   33

Investigation 4

shape, triangle, square, hexagon, trapezoid, diamond

1. Using pattern blocks, point out that different blocks have different shapes. Draw or trace five shapes on the board—triangle, square, hexagon, trapezoid, and diamond—and identify each.
2. Let students each take a handful of pattern blocks. Ask students to hold up one of each shape as you name it. Point to the shapes on the board for the first round; for a second round, erase the shapes you have drawn and just call for the pieces by name, using action commands.

   Find a triangle.

   Give a hexagon to someone else.

   Put a trapezoid in the middle of the table.

   Show me all the blue diamonds.

Blackline Masters

________________________, 19 ____

Dear Family,

We are beginning a unit called *Mathematical Thinking at Grade 4.* This unit will help your child get used to solving problems that take considerable time, thought, and discussion. While solving these problems, your child will be using materials like coins, cubes, and calculators, and will be writing, drawing, and talking about how to do the problems. Emphasis during this unit will be on thinking hard and reasoning carefully to solve mathematical problems.

During this unit your child will estimate how many hundreds are in a group of objects. The class will also be figuring out ways to count coins efficiently, creating symmetrical designs, and looking for patterns in their environment. In what we call *working with data,* the class will be collecting and organizing information about themselves. Your child will have a math folder or journal for keeping track of work.

While our class is working on this unit and throughout the year, you can help in several ways:

- Your child will have assignments to work on at home. Sometimes they will involve your participation. For example, your child will be teaching you several games that can be played cooperatively. Later in the unit, your child will be asked to look for patterns in magazines or newspapers or on pieces of fabric, gift wrap, or wallpaper.
- During this unit, students will be solving problems involving money. When working at home, it would help them to have an assortment of coins to work with. You can also involve your child in solving problems about money: How much will two items cost together? How much change will you get for $1.00? What is the total value of all the coins in a purse or pocket?

One of the most important things you can do is to show genuine interest in the ways your child solves problems, even if they are different from your own. We are looking forward to an exciting few weeks as we create a mathematical community in our classroom.

Sincerely,

Name Date

Student Sheet 1

How Many Cubes in Each Object?

Name of maker	Name of object	Estimate	Count

Tell about what you enjoyed in math class today.

Suppose someone was trying to estimate how many cubes were in an object. What advice would you give that person?

Name Date

Student Sheet 2

Making Hundreds

1. Make a drawing to show how you grouped 100 cubes. (If you arranged 100 in more than one way, show both ways.)

2. How do you know you have exactly 100 cubes?

3. What other ways could you have arranged 100 cubes?

Name Date

Student Sheet 3

How Many Cubes in the Class?

Number of cubes in each group	About how many hundreds altogether? Estimate.	Add the numbers to find the exact total.
	Show or explain how you got your answer.	Is the total close to the number of hundreds you estimated? If they are very different, try to find out why.

Name Date

Student Sheet 4

A Design with 100 Squares

On graph paper, make a design using exactly 100 squares.

Finish the sentence below:

I know my design is 100 squares because ______________________

____

____

____

____

____

Name Date

Student Sheet 5

How Many Hundreds? How Many Altogether?

These are the numbers of cubes that groups in two classes counted.

Number of cubes in each group	About how many hundreds altogether? Estimate.	Add the numbers to find the exact total.
46 22 74 **51 98** **113**	100 or 200 or 300 or 400? How did you decide?	
131 271 **197 204**	600 or 700 or 800 or 900? How did you decide?	

Name Date

Student Sheet 6

Problems for Close to 100

Pretend you are dealt these hands in the game Close to 100. What numbers would you make to get sums as close to 100 as possible?

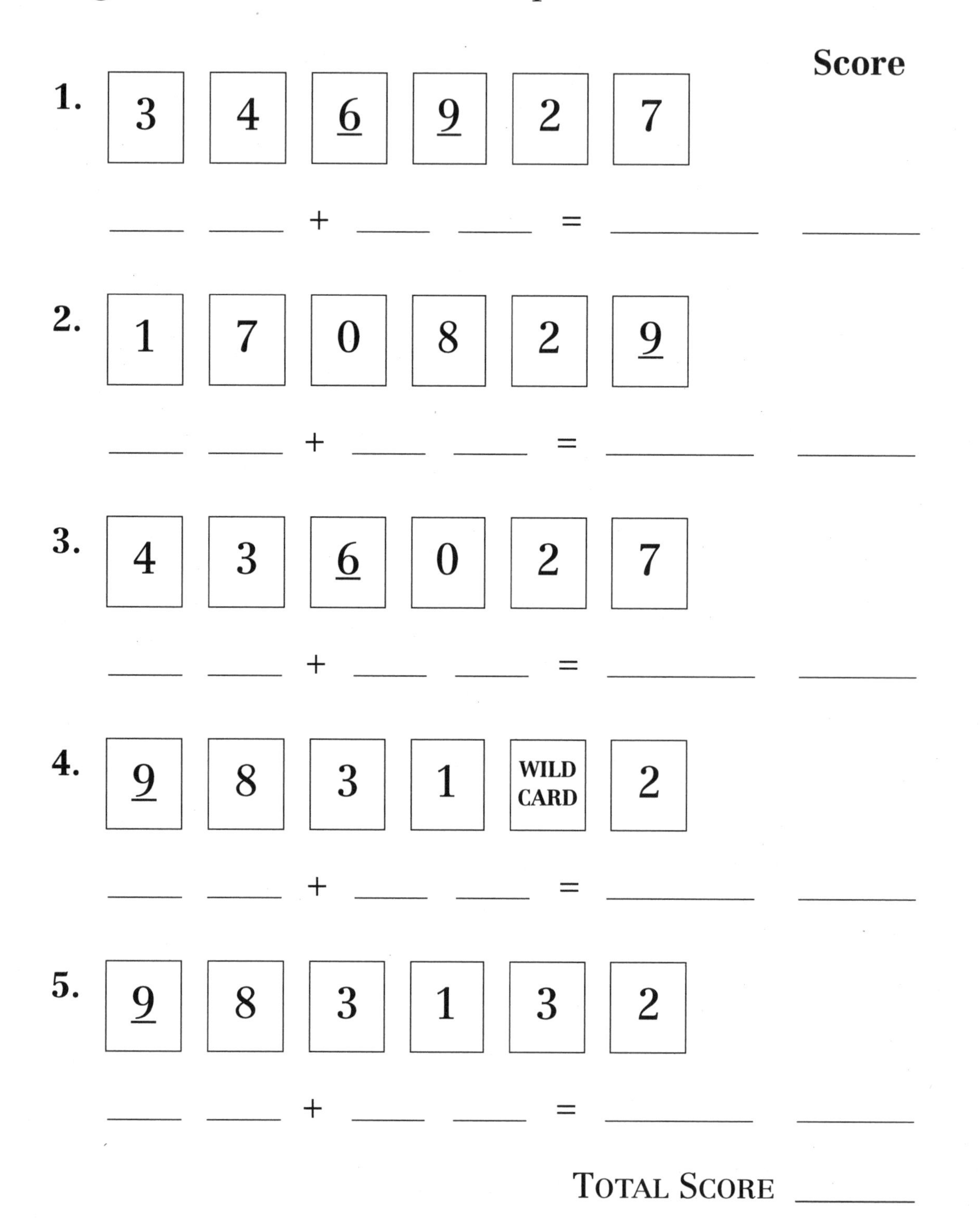

TOTAL SCORE ______

CHART FOR HOW MANY CUBES?

Group	Estimates	Counts

Materials

- One deck of Numeral Cards
- Close to 100 Score Sheet for each player

Players: 1, 2, or 3

How to Play

1. Deal out six Numeral Cards to each player.
2. Use any four of your cards to make two numbers. For example, a 6 and a 5 could make either 56 or 65. Wild Cards can be used as any numeral. Try to make numbers that, when added, give you a total that is close to 100.
3. Write these two numbers and their total on the Close to 100 Score Sheet. For example: $42 + 56 = 98$.
4. Find your score. Your score is the difference between your total and 100. For example, if your total is 98, your score is 2. If your total is 105, your score is 5.
5. Put the cards you used in a discard pile. Keep the two cards you didn't use for the next round.
6. For the next round, deal four new cards to each player. Make more numbers that come close to 100. When you run out of cards, mix up the discard pile and use them again.
7. Five rounds make one game. Total your scores for the five rounds. LOWEST score wins!

Scoring Variation

Write the score with plus and minus signs to show the direction of your total away from 100. For example: If your total is 98, your score is –2. If your total is 105, your score is +5. The total of these two scores would be +3. Your goal is to get a total score for five rounds that is close to 0.

ONE-CENTIMETER GRAPH PAPER

Name ________________________________ Date

Student Sheet 7

Ways to Count Money

Below are three handfuls of coins. Find the total value of each. Try it in two or three different ways.

- Try it mentally or with coins.
- Try it with a calculator.
- Try it with paper and pencil.

1. 2 quarters
3 dimes
2 nickels
3 pennies
Total value: __________

2. 1 half dollar
1 quarter
1 dime
7 nickels
Total value: __________

3. 3 quarters
4 pennies
5 nickels
3 dimes
Total value: __________

4. What is the best way to find the total value of a handful of coins? What advice would you give to a friend about this?

____

____

____

____

____

COIN CARDS (page 1 of 4)

LIBERTY
IN GOD WE TRUST
1994
UNITED STATES OF AMERICA
E PLURIBUS UNUM
QUARTER DOLLAR
FIVE CENTS
MONTICELLO
ONE CENT

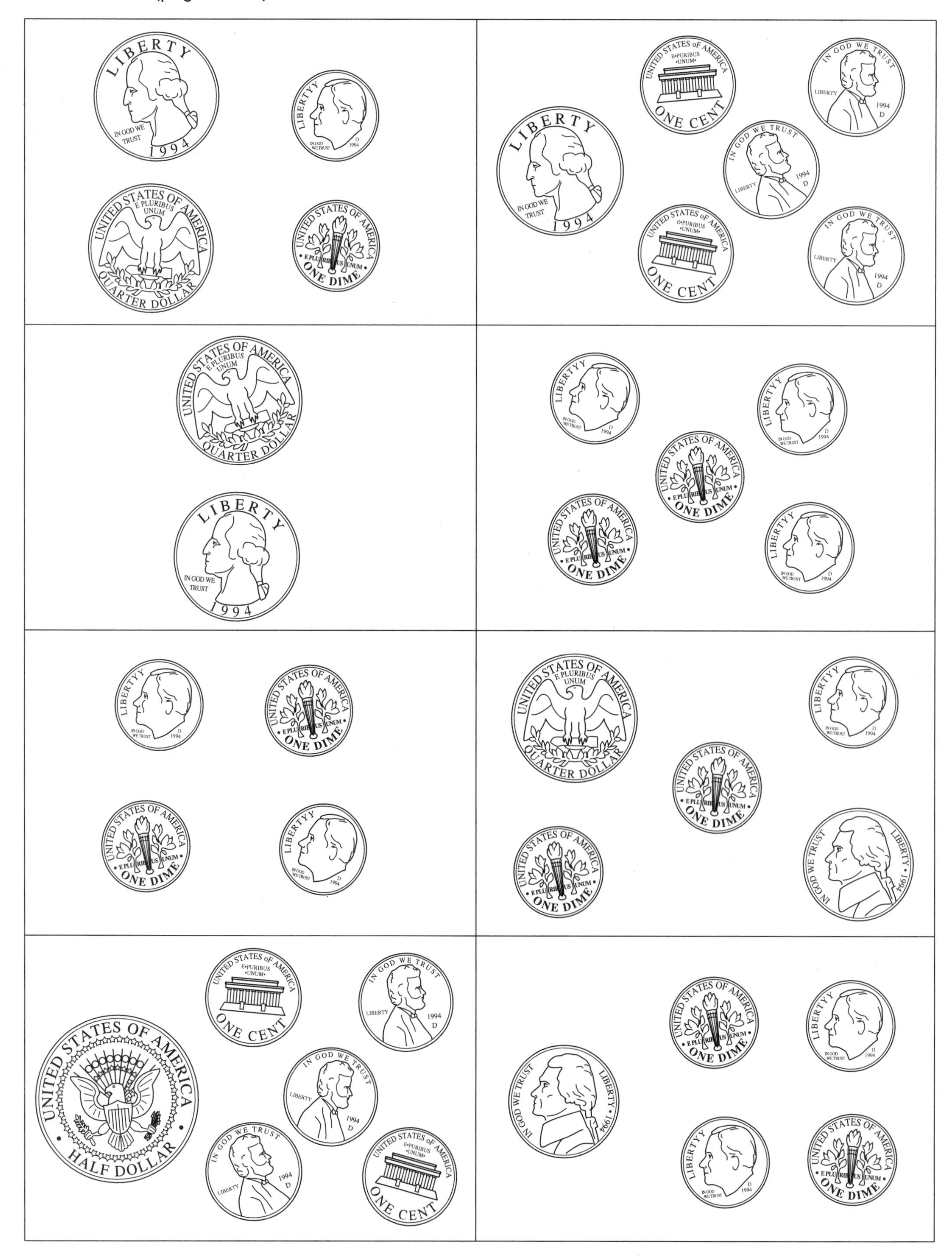

Materials

- One deck of Coin Cards

Players: 2 or 3

How to Play

1. Deal out 8 Coin Cards face up. Put the rest of the pack in a pile face down.
2. Take turns finding pairs of cards that equal one dollar.
3. If everyone agrees that no combinations of the 8 cards make one dollar, shuffle all 8 back in the pack and deal 8 new cards.
4. At the end of your turn, draw new cards from the pack to replace the cards that you used. The next player should have 8 cards to choose from.
5. The goal is to collect as many pairs of cards (dollars) as possible.
6. The game ends when all the cards have been paired.

Scoring Variation

Make combinations of cards that equal any whole number of dollars. For example, a player could take three cards with 50¢, 70¢, and 80¢ for a total of $2.00. You may decide to score by number of dollars collected or by number of cards collected.

Materials

- One dollar in real coins:
  2 quarters, 3 dimes, 3 nickels, and 5 pennies
- A small paper bag you can *reach* into but not *see* into. Put the coins in the bag.

Players: 2 or 3 (can also be a solitaire game)

How to Play

1. Players agree on a sum of money less than $1.00 that one player will try to pick from the bag. Some easy amounts require only 1 or 2 coins; for example, 10¢ or 35¢. More difficult amounts require more coins; for example, 23¢, 47¢, 66¢, or 92¢.
2. Reach into the bag and take out one coin at a time until you have the target amount of money. If you take a coin that will not help you make the target amount, put it back.
3. When all players agree that the target amount has been picked out, return the coins to the bag.
4. Choose a different amount of money, and start again. Take turns.

Variation

After one player picks out the right amount, the next player tries to make the same amount of money with different coins.

If no other way is possible, that player picks out the same selection of coins the first player chose.

Name Date

Student Sheet 8

300 Chart (page 1 of 3)

1	2	3							10
11					16				20
		23			26				
		33				37			
41	42								
		53	54						60
				65					
	72								
81									90
						97			100

Trim along bottom of chart. Tape to next page.

Name ______________________________ Date

Student Sheet 8

300 Chart (page 2 of 3)

Tape bottom of page 1 along here.

			104						
	112	113							120
				125		127			
131									140
		143	144						150
				155	156				
161									170
				175					
	182	183							190
								199	

Trim along bottom of chart. Tape to next page.

Name Date

Student Sheet 8

300 Chart (page 3 of 3)

Tape bottom of page 2 along here.

		203				207			
211			214						220
				225					
	232								240
		243	244						
251					256				
			264						
271									
								289	290
						297			

Name Date

Student Sheet 9

Related Problem Sets (page 1 of 6)

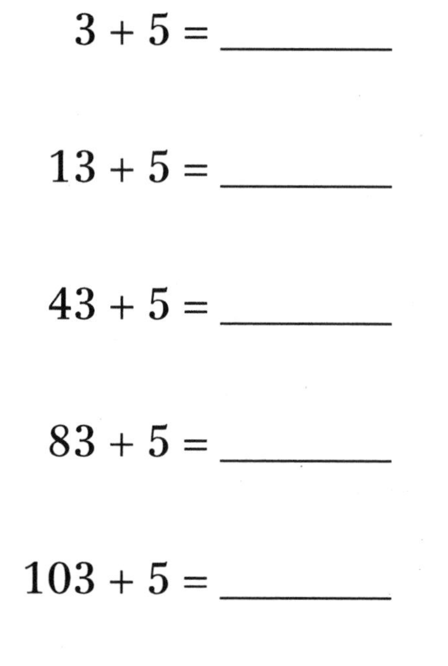

$3 + 5 =$ ________

$13 + 5 =$ ________

$43 + 5 =$ ________

$83 + 5 =$ ________

$103 + 5 =$ ________

How I solved this set

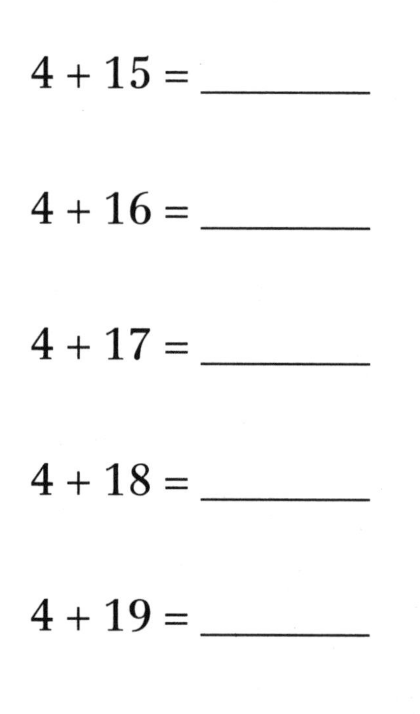

$4 + 15 =$ ________

$4 + 16 =$ ________

$4 + 17 =$ ________

$4 + 18 =$ ________

$4 + 19 =$ ________

How I solved this set

Name ______________________________ Date

Related Problem Sets (page 2 of 6)

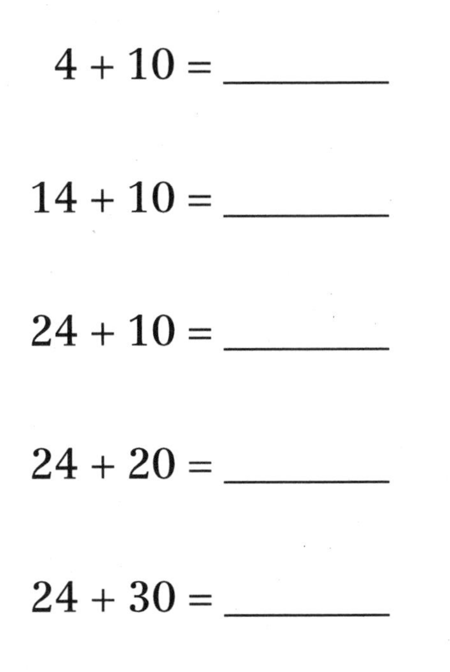

$4 + 10 =$ ________

$14 + 10 =$ ________

$24 + 10 =$ ________

$24 + 20 =$ ________

$24 + 30 =$ ________

How I solved this set

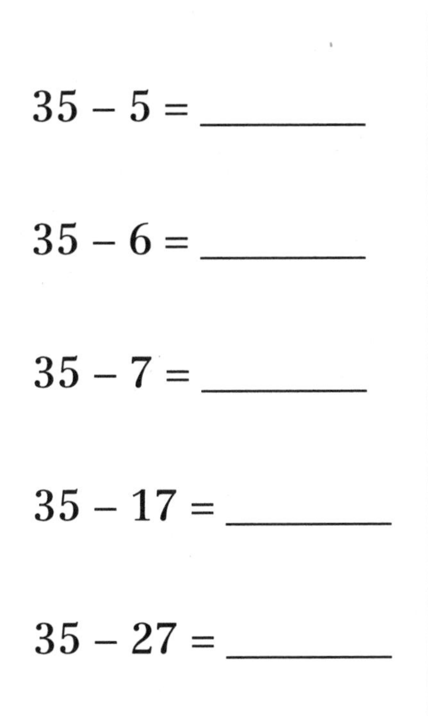

$35 - 5 =$ ________

$35 - 6 =$ ________

$35 - 7 =$ ________

$35 - 17 =$ ________

$35 - 27 =$ ________

How I solved this set

Name Date

Student Sheet 9

Related Problem Sets (page 3 of 6)

32 + 10 = ________

32 + 20 = ________

32 + 40 = ________

32 + 70 = ________

32 + 80 = ________

How I solved this set

---

80 + ________ = 100

180 + ________ = 200

280 + ________ = 300

180 + ________ = 300

How I solved this set

Name ______________________ Date ______

Related Problem Sets (page 4 of 6)

1 quarter = _______ pennies

2 quarters = _______ pennies

3 quarters = _______ pennies

4 quarters = _______ pennies

5 quarters = _______ pennies

How I solved this set

---

1 dime = _______ nickels

2 dimes = _______ nickels

5 dimes = _______ nickels

10 dimes = _______ nickels

20 dimes = _______ nickels

How I solved this set

Name ________________ Date

Related Problem Sets (page 5 of 6)

1 quarter = ________ nickels

2 quarters = ________ nickels

3 quarters = ________ nickels

4 quarters = ________ nickels

8 quarters = ________ nickels

How I solved this set

---

1 dollar = ________ pennies

1 dollar = ________ nickels

2 dollars = ________ nickels

1 dollar = ________ dimes

3 dollars = ________ dimes

How I solved this set

Name Date

Related Problem Sets (page 6 of 6)

Write the total value for each of these groups of coins.

2 nickels =

3 nickels =

5 nickels =

6 nickels =

7 nickels =

How I solved this set

---

1 quarter and 1 nickel =

1 quarter and 1 dime =

2 quarters and 1 dime =

3 quarters =

3 quarters and 1 nickel =

How I solved this set

Name Date

Student Sheet 10

Numbers and Money (page 1 of 2)

1. Here are 6 Numeral Cards:

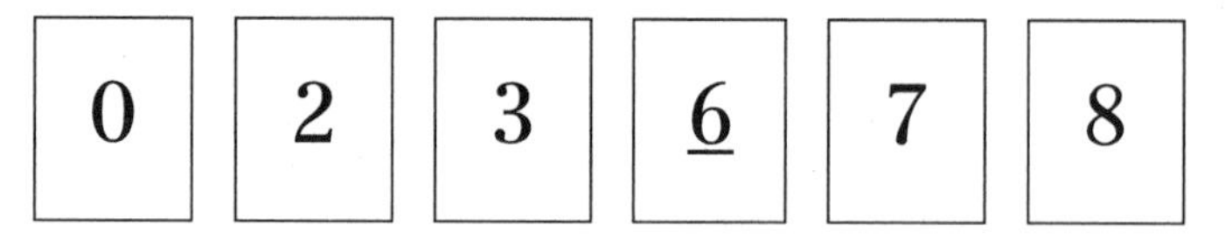

Use four of the numbers. Make two 2-digit numbers that you can add to make a number close to 100. Can you get 100 exactly? How close can you get?

2. Write the answers to these counting problems in the blanks:

a. Start with 58. Count up by 10's by adding 10 each time.

58 ____ ____ ____ ____ ____ ____ (+ 10 each time)

b. Start with 4. Count up by 20's.

4 ____ ____ ____ ____ ____ ____ (+ 20 each time)

c. Start with 137. Count backward by 10's.

137 ____ ____ ____ ____ ____ ____ (– 10 each time)

d. What advice would you give to someone who was trying to count by 10's and 20's?

Name Date

Student Sheet 10

Numbers and Money (page 2 of 2)

3. Count and make groups of coins.

a. What is the total value of these coins?

b. What is the total value of 3 quarters, 2 dimes, 2 nickels, and 2 pennies?

c. Show what coins you could use to make 63¢.

d. Show a different way to make 63¢.

Materials

- 101 to 200 Bingo Board
- One deck of Numeral Cards
- One deck of Tens Cards
- Colored pencil, crayon, or marker

Players: 2 or 3

How to Play

1. Each player takes a 1 from the Numeral Card deck and keeps this card throughout the game.
2. Shuffle the two decks of cards. Place each deck face down on the table.
3. Players use just one Bingo Board. You will take turns and work together to get a Bingo.
4. To determine a play, draw two Numeral Cards and one Tens Card. Arrange the 1 and the two other numerals to make a number between 100 and 199. Then add or subtract the number on your Tens Card. Circle the resulting number on the 101 to 200 Bingo Board.
5. Wild Cards in the Numeral Card deck can be used as any numeral from 0 through 9. Wild Cards in the Tens Card deck can be used as + or – any multiple of 10 from 10 through 70.
6. Some combinations cannot land on the 101 to 200 Bingo Board at all. Make up your own rules about what to do when this happens. (For example, a player could take another turn, or the Tens Card could be *either* added or subtracted in this instance.)
7. The goal is for the players together to circle five adjacent numbers in a row, in a column, or on a diagonal. Five circled numbers is a Bingo.

101 TO 200 BINGO BOARD

101	102	103	104	105	106	107	108	109	110
111	112	113	114	115	116	117	118	119	120
121	122	123	124	125	126	127	128	129	130
131	132	133	134	135	136	137	138	139	140
141	142	143	144	145	146	147	148	149	150
151	152	153	154	155	156	157	158	159	160
161	162	163	164	165	166	167	168	169	170
171	172	173	174	175	176	177	178	179	180
181	182	183	184	185	186	187	188	189	190
191	192	193	194	195	196	197	198	199	200

+10	**+10**	**+10**	**+10**
+20	**+20**	**+20**	**+20**
+30	**+30**	**+30**	**+40**
+40	**+50**	**+50**	**+60**
+70	**WILD CARD**	**WILD CARD**	**WILD CARD**

-10	-10	-10	-10
-20	-20	-20	-20
-30	-30	-30	-40
-50	-50	-50	-60
-70	WILD CARD	WILD CARD	WILD CARD

Name Date

Student Sheet 11

Mirror Symmetry and Rotational Symmetry

One of these designs has mirror symmetry.
The other design has rotational symmetry.
Cut out the designs. Try turning and folding them.
Can you tell which design has mirror symmetry?
Which design has rotational symmetry?

Draw the line of symmetry in the design with mirror symmetry.
Put a dot in the center of the design with rotational symmetry.

Color the designs to show the symmetry.

Name Date

Student Sheet 12

Multiple Lines of Symmetry

Use geoboard dot paper, triangle paper, or plain paper to make the following designs:

Make one design with only one line of symmetry.

Make one design with two lines of symmetry.

Make one design with four lines of symmetry.

Label the lines of symmetry in your designs.

SHADED GEOBOARD DESIGN

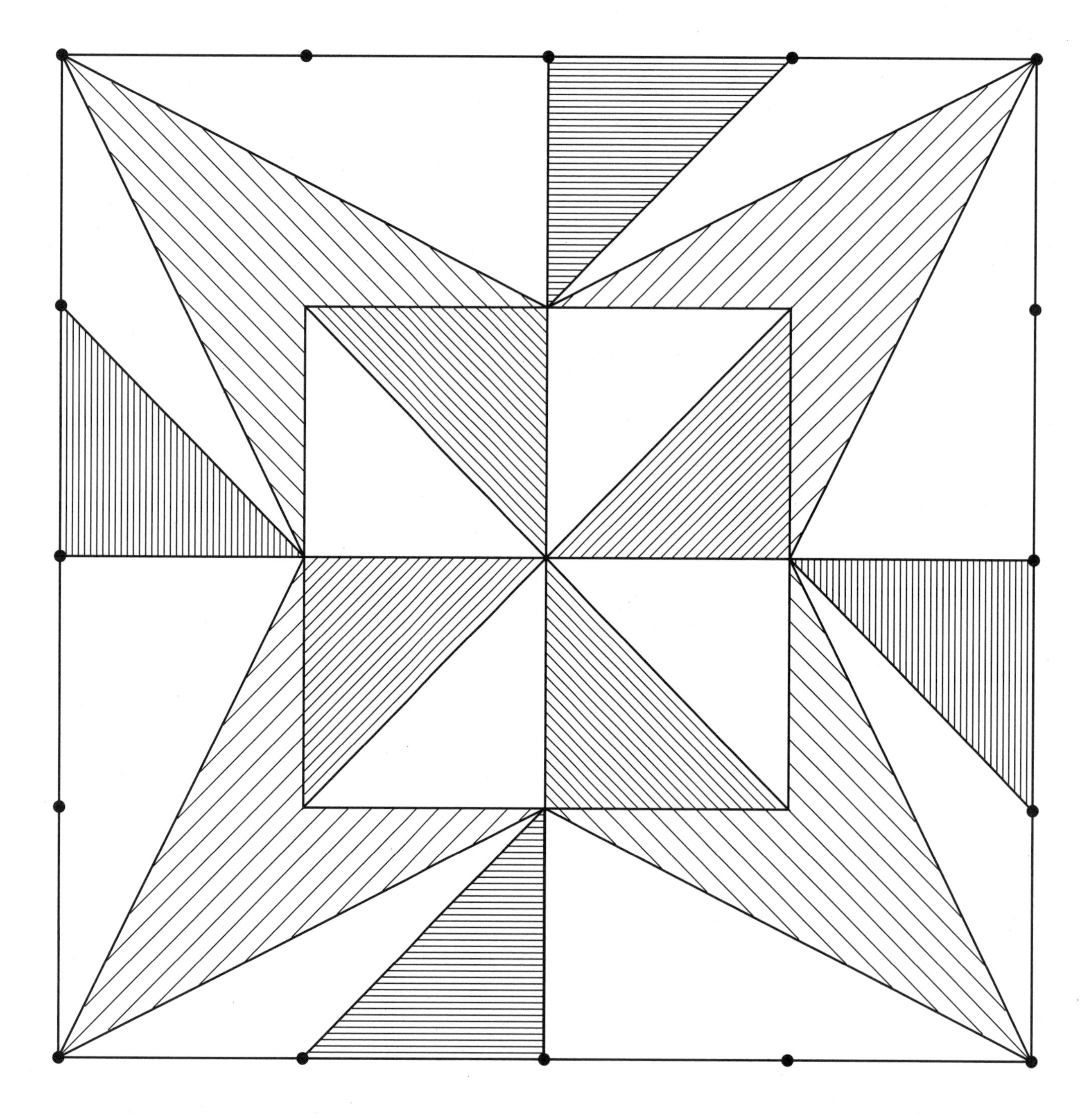

GEOBOARD DOT PAPER

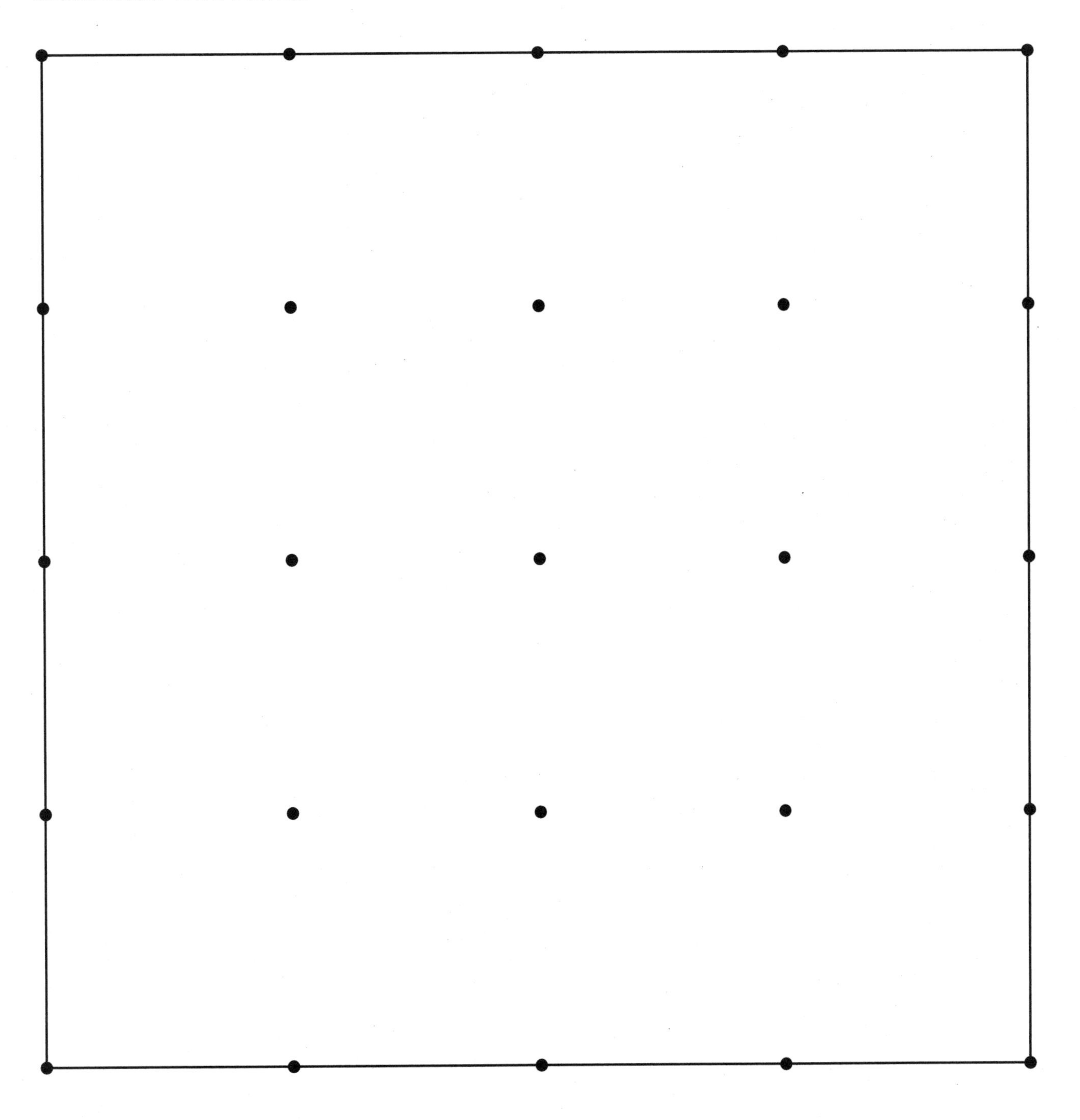

Name ____

GAME 1 Score

Round 1: ____ ____ + ____ ____ = ________ ______

Round 2: ____ ____ + ____ ____ = ________ ______

Round 3: ____ ____ + ____ ____ = ________ ______

Round 4: ____ ____ + ____ ____ = ________ ______

Round 5: ____ ____ + ____ ____ = ________ ______

TOTAL SCORE ______

Name ____

GAME 2 Score

Round 1: ____ ____ + ____ ____ = ________ ______

Round 2: ____ ____ + ____ ____ = ________ ______

Round 3: ____ ____ + ____ ____ = ________ ______

Round 4: ____ ____ + ____ ____ = ________ ______

Round 5: ____ ____ + ____ ____ = ________ ______

TOTAL SCORE ______

NUMERAL CARDS (page 1 of 3)

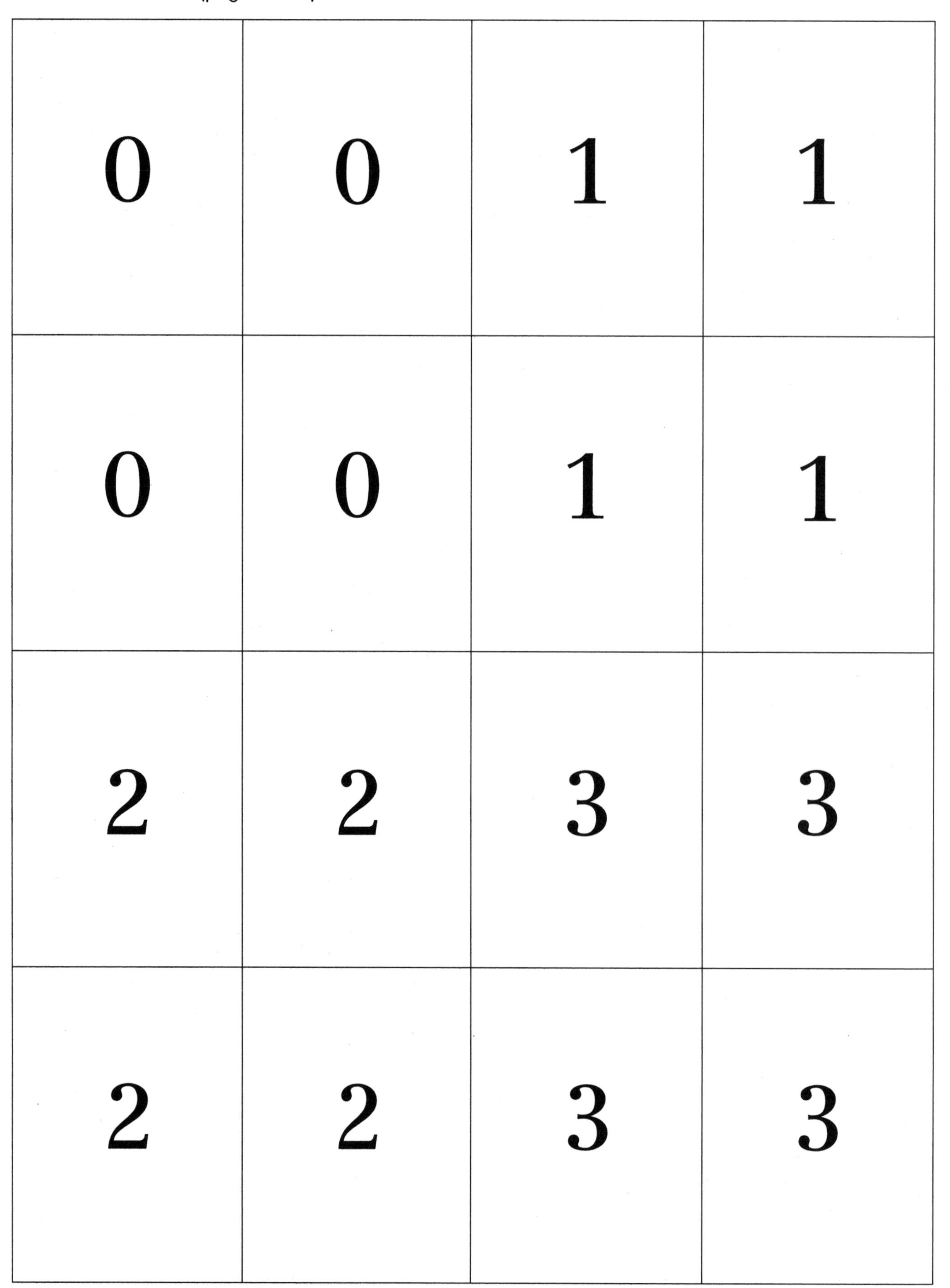

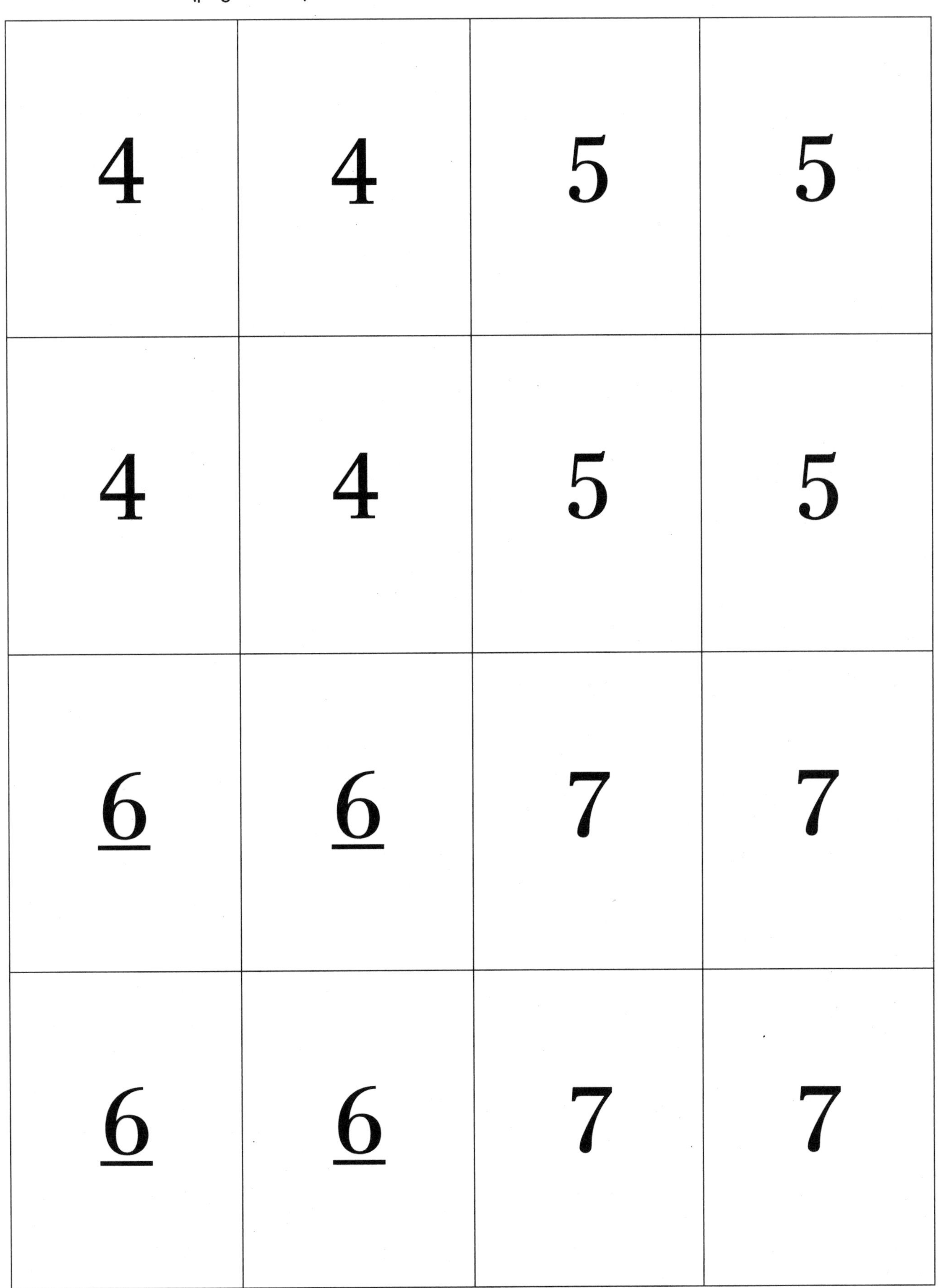
4
4
5
5
4
4
5
5
6
6
7
7
6
6
7
7

8	8	9	9
8	8	9	9
WILD CARD	WILD CARD		
WILD CARD	WILD CARD		

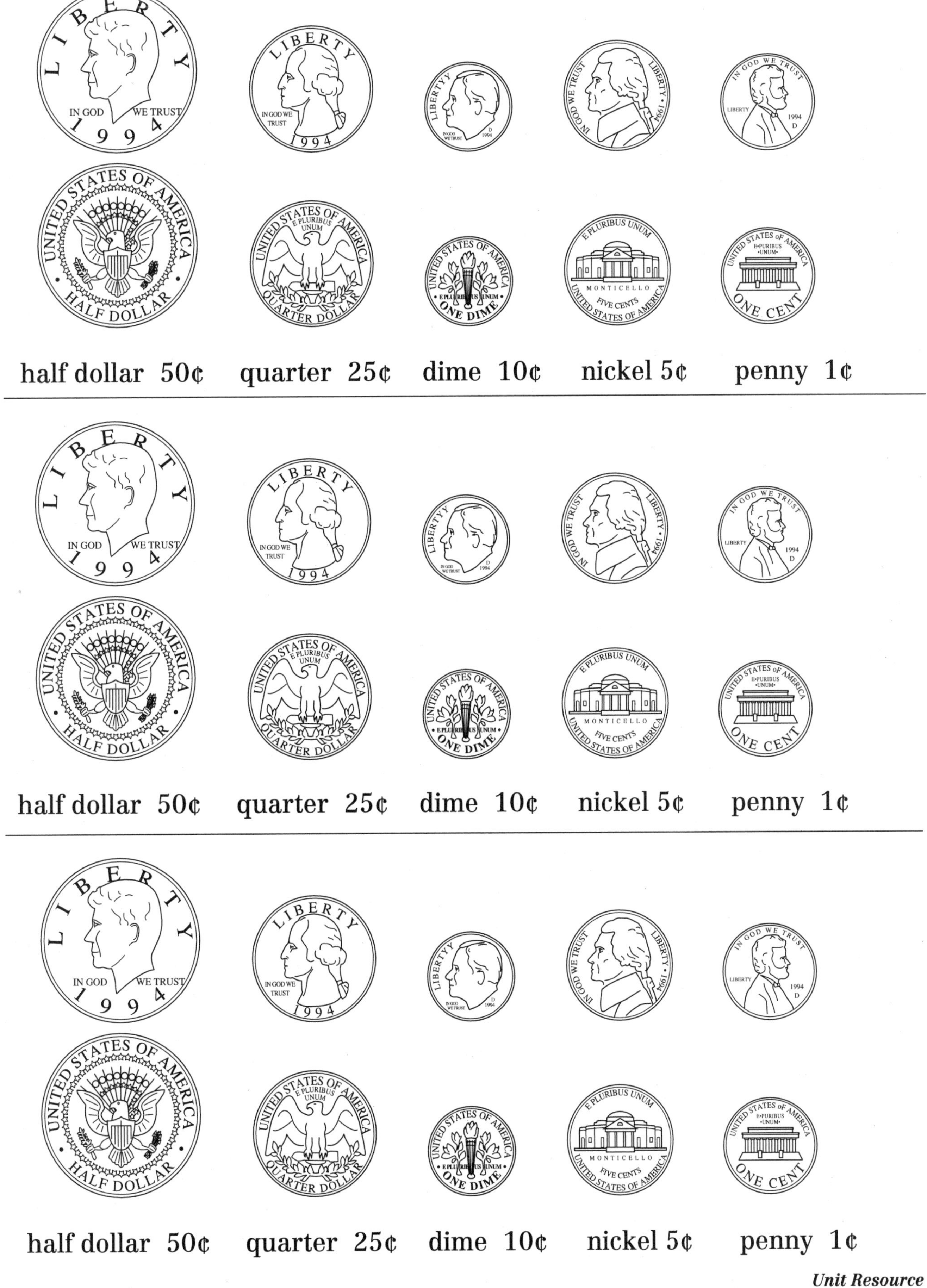
half dollar 50¢
quarter 25¢
dime 10¢
nickel 5¢
penny 1¢
half dollar 50¢
quarter 25¢
dime 10¢
nickel 5¢
penny 1¢
half dollar 50¢
quarter 25¢
dime 10¢
nickel 5¢
penny 1¢

Name ________________________________

Activity Choice	✔ When Finished
1. ______________________	☐
2. ______________________	☐
3. ______________________	☐
4. ______________________	☐
5. ______________________	☐
6. ______________________	☐
7. ______________________	☐
8. ______________________	☐

Practice Pages

This optional section provides homework ideas for teachers who want or need to give more homework than is assigned to accompany the activities in this unit. The problems included here provide additional practice in learning about number relationships and in solving computation and number problems. For number units, you may want to use some of these if your students need more work in these areas or if you want to assign daily homework. For other units, you can use these problems so that students can continue to work on developing number and computation sense while they are focusing on other mathematical content in class. We recommend that you introduce activities in class before assigning related problems for homework.

Solving Problems in Two Ways Solving problems in two ways is emphasized throughout the *Investigations* fourth grade curriculum. Here, we provide four sheets of problems that students solve in two different ways. Problems may be addition, subtraction, multiplication, or division. Students record each way they solved the problem. We recommend you give students an opportunity to share a variety of strategies for solving problems before you assign this homework.

Story Problems Story problems at various levels of difficulty are used throughout the *Investigations* curriculum. The three story problem sheets provided here help students review and maintain skills that have already been taught. You can also make up other problems in this format, using numbers and contexts that are appropriate for your students. Students solve the problems and then record their strategies.

Name Date

Practice Page A

Solve this problem in two different ways, and write about how you solved it:

90 + 45 =

Here is the first way I solved it:

Here is the second way I solved it:

Name Date

Practice Page B

Solve this problem in two different ways, and write about how you solved it:

75 − 16 =

Here is the first way I solved it:

Here is the second way I solved it:

Name Date

Practice Page C

Solve this problem in two different ways, and write about how you solved it:

$$11 \times 7 =$$

Here is the first way I solved it:

Here is the second way I solved it:

Name Date

Practice Page D

Solve this problem in two different ways, and write about how you solved it:

$32 \div 4 =$

Here is the first way I solved it:

Here is the second way I solved it:

Name Date

Practice Page E

The football team scored 4 touchdowns. They got 7 points for each touchdown. How many points do they have?

Show how you solved this problem. You can use numbers, words, or pictures.

Name Date

Practice Page F

James and I made chocolate chip cookies to sell. We made 2 batches. Each batch makes 24 cookies. How many cookies can we sell?

Show how you solved this problem. You can use numbers, words, or pictures.

Name Date

Practice Page G

There are 6 people in my family. We bought 2 pizzas. Each pizza has 12 slices. If we share equally, how many slices of pizza will each person get?

Show how you solved this problem. You can use numbers, words, or pictures.